Alim-un- Nisa
Sajila Hina
Hamood-ur- Rehman

The Truth on Artificial Sweeteners

Alim-un- Nisa
Sajila Hina
Hamood-ur- Rehman

The Truth on Artificial Sweeteners

Noor Publishing

Cover image: www.ingimage.com

Publisher:
Noor Publishing
is a trademark of
Dodo Books Indian Ocean Ltd., member of the OmniScriptum S.R.L Publishing group
str. A.Russo 15, of. 61, Chisinau-2068, Republic of Moldova Europe
Printed at: see last page
ISBN: 978-620-3-85776-4

Food & Biotechnology Research Centre.

PCSIR Laboratories Complex, Ferozepur Road,

Lahore-54600, Pakistan

Table of Contents

ARTIFICIAL SWEETENERS

INTRODUCTION:

Taste perception is an important attribute of human beings and the human tongue can detect four basic flavors, salt, sour, bitter and sweet. It is natural that humans are mostly drawn to sweet and sweet taste always fascinated almost everyone. Whether it may be the eating of natural sweet foods such as fruits, honey or the processed foods with added sugar, it is always appealing. Sweetened foods supplies more ready energy and also provide a pleasure. As the food industry progresses there arouse a need for the addition of sugar in the processed food to have a variety of delicious sweet food products. This paved the way of extracting sugar from sugar cane, beets and corn and now the sugar consumption is almost 120 million tons per year with a still expanding market. There are various sweetening agent present naturally that enhance taste of everyday foods and have great nutritional importance. Sweeteners are either monosaccharide or disaccharide. Natural sweeteners are extracted through honey, maple syrup, Molasses sans sucrose (Spillane, 2006). One of essential natural sweetener used in human diet is table sugar. Harmful effects of table sugar on health provide a path to sugar alternatives. Many chronic diseases such as coronary heart disease, obesity, diabetes, and hypertension were linked to excessive consumption of sugar. Food industries also synthesize some chemicals that induce sweet taste but have no nutritional importance but are sweeter than natural sweetener. These sugar alternatives are known as artificial sweeteners (AS) or non-caloric artificial sweeteners (NCAS) and are abundant component of sugar free diet. Low-calorie sweeteners are ingredients added to foods and yogurt and sugar-free pudding.

They also play an important role in a weight management program that includes both good nutrition choices and physical activity. Artificial sweeteners are also known as intense sweeteners because they are many times sweeter than regular sugar.

INTENSE SWEETENERS:

Two types of intense sweeteners are available:

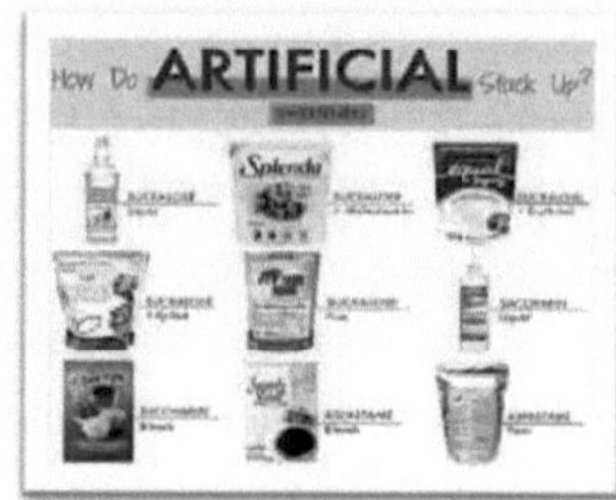

1. **Natural sweeteners of plant origin**
2. **Synthetic sweeteners.**

The sweeteners from ordinary sources with potential for commercial use include perillaldehyde, stevioside, rabaudioside, glycyrrhizin, osladin, thaumatins, and monellin. Low calorie sweeteners are sugar replacements that have zero calories and do not raise blood glucose levels through eating them, which makes them a preferable choice for diabetic people over sugar. The main reason behind use of these AS is because they are sweeter then table sugar, low in calories and cost, reduce weight and stabilize blood sugar level. Due to low cost and calories these NCAS are commonly used in diet sodas, cereals sugar free desserts and also recommended for those patients that are suffering from type II diabetes. Five artificial sweeteners Acesulfame K, Aspartame, Neotame, Saccharin, and Sucralose are approved for use in the U.S. All are chemically manufactured molecules that do not be present in nature. Artificial sweeteners are used in one of two ways. They may be used directly in commercially processed foods, or they are mixed with one or more starch based sweeteners before deal to consumers. Millions of health conscious people worldwide consider sugar alternatives a healthy substance which is not true.

TYPES OF ARTIFICIAL SWEETENERS:

Artificial sweeteners are regulated by the U.S. Food and Drug Administration (FDA). The FDA, like the National Cancer Institute (NCI), is an agency of the Department of Health and Human Services. The FDA regulates food, drugs, medical devices, cosmetics, biologics, and radiation-emitting products. The Food Additives Amendment to the Food, Drug, and Cosmetic Act, which was passed by Congress in 1958, requires the FDA to approve food additives, including artificial sweeteners, before they can be made available for sale in the United States. However, this legislation does not apply to products that are "generally recognized as safe." Such products do not require FDA approval before being marketed.Currently, the FDA has approved Six artificial sweeteners for consumption: acesulfame-K, aspartame, neotame, saccharin, and sucralose and advantame (Table I).In addition, Stevia is also "generally recognized as safe" (GRAS) by FDA whereas Cyclamate is banned since 1969 for and in 1970 it is completely banned for all kind of food and drug products. Artificial sweeteners are known by several names, which include: low-calorie sweeteners, high intensity sweeteners, non-sucrose sweeteners, intense sweeteners, non-nutritive sweeteners, sugar substitutes, and sugar-free sweeteners (Duffy and Sigman-Grant, 2004).

Table No: 1

Types of artificial sweeteners

Chemical Name	Trade Names	Sweetness (compared with sucrose)	Acceptable Daily Intake (ADI)	Structure
Saccharin ($C_7H_5NO_3S$)	Sweet N' Low	300×	5 mg/kg	

Aspartame ($C_{14}H_{18}N_2O_5$)	NutraSweet, Equal	160–220×	50 mg/kg	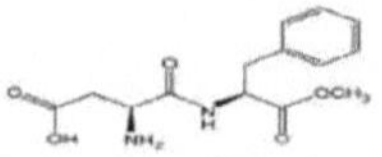
Acesulfame-Potassium ($C_4H_4KNO_4S$)	Sunett, Sweet & Safe, Sweet One	200×	15 mg/kg	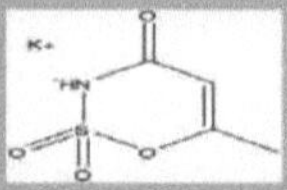
Sucralose ($C_{12}H_{19}Cl_3O_8$)	Splenda	600×	5 mg/kg	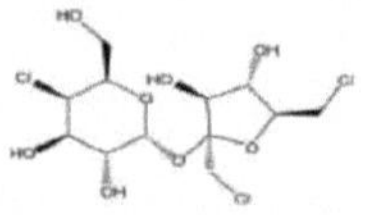
Neotame ($C_2OH_3ON_2O_5$)	Made by NutraSweet	7 000–13 000×	0.10 mg/kg	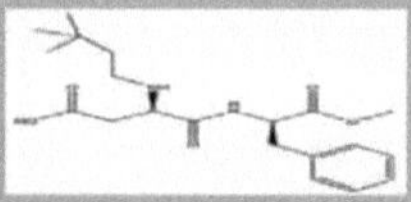
Advantame ($C_{24}H_{30}N_2O_7$)	Ajisweet	20,000 ×	5 mg/kg	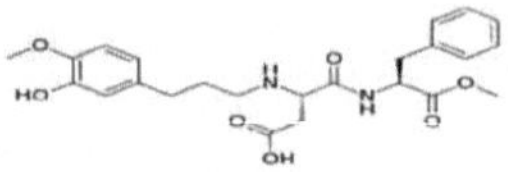

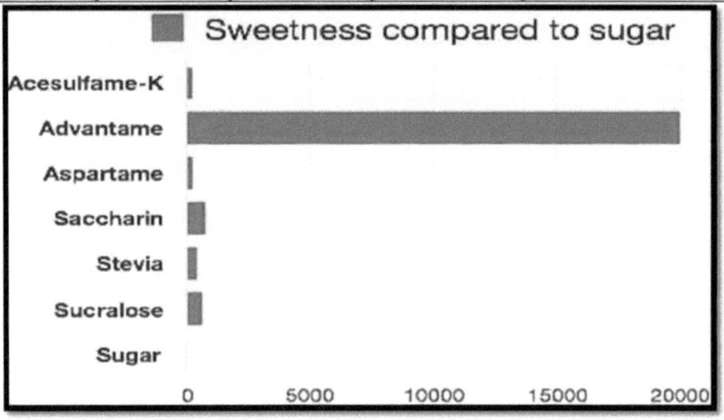

Types of High Intensity Sweetener	Examples
Synthetic chemicals	-Saccharine. - Acesulfame-K - Cyclamate
Semi Synthetic Sugar	-Sucralose
Semi-Synthetic Peptides	-Aspartame - Neotame - Advantame - Alitame
Natural extracts	-Steviol glycosides - Mogrosides - Thaumatin - Brazzein

IMPORTANT FACTORS FOR SWEETENERS PERCEPTION:

The strength of taste development and perception depends on physical and chemical composition of medium in which sweetener is dispensed, sweeteners concentration, temperature and the pH of medium.

SOURCES:

Sweetness can be enhanced by making combination of more than two sweetening agents. Different sweeteners have different intensity, duration and after-taste, solubility and stability at different pH and temperature, so making combination will improve taste. NCAS when used in combination can increase sweetness; to reduce this sweet taste polyols (carbohydrates that are not sweet) are added. This combination of NCAS is preferred by companies because these NAS doesn't react with each other so they are safe to use. The caloric values for the polyols approved by FDA are from 1.6 to 3.0 depending type of polyols.

Combination of saccharin and aspartame is present in soft drinks, aspartame and acesulfame potassium in combination with sucralose are present in bottled drink. Combination of polyols and sweeteners is abundantly used in chewing-gums(Nabors, 2012).

USES FOR ARTIFICIAL SWEETENERS

Artificial sweeteners are attractive alternatives to sugar because they add virtually no calories to your diet. In addition, you need only a fraction compared with the amount of sugar you would normally use for sweetness. Artificial sweeteners are widely used in processed foods, including baked goods, soft drinks, powdered drink mixes, candy, puddings, canned foods, jams and jellies, dairy products, and scores of other foods and beverages.

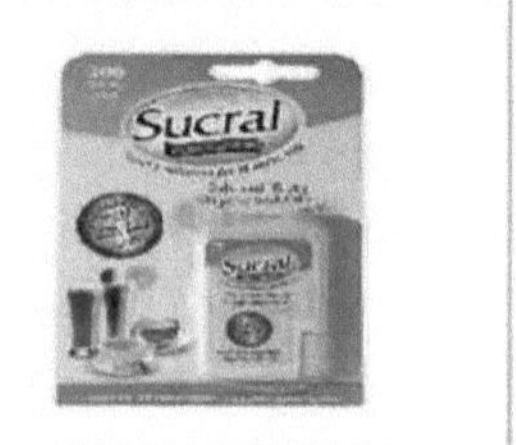

Carbonated and other beverages
(Juices, iced tea, flavored waters, etc.)

- Cereals
- Chewing gums
- Dairy product (low-fat flavored milk, light yogurt, etc.)
- Canned fruits
- Desserts
  (Light pudding, light ice-cream, popsicles, etc.)
- Nutrition bars
- Syrups and fruit spreads
- Tabletop sweeteners

➢ Why Artificial Sweeteners are Used:

Artificial sweeteners are used for being giving sweetening and as alternative of sweetness. Sugar and sugar alcohols are used for sweetening purposes. Basically, sugar alcohols are carbohydrate structures. Sugar alcohols are beneficial substitute for sweetness purposes.

➢ Artificial Sweeteners Brand Names

Here are some common brands of artificial sweeteners available in the market

Sweetener Name	Brand Names Found in Stores
Acesulfame Potassium	Sunett
	Sweet One
Aspartame	Nutrasweet

	Equal
Saccharin	Sweet 'N Low
	Sweet Twin
	Sugar Twin
Sucralose	Splenda
Stevia/Rebaudioside A	A Sweet Leaf
	Sun Crystals
	Steviva
	Truvia
	Purevia

SACCHARIN(O-SULFABENZAMIDE;2,3-DIHYDRO-3 OXOBENZISOSULFONAZOLE):

Saccharin is available in three forms depending upon physical properties as well as ion added on it i.e. sodium, calcium and acid saccharin. It is white crystalline powder. Sodium saccharin has high solubility in water and used in food industry as flavoring agent., acid saccharin is slightly soluble in water and has its use in tobacco and toothpaste industries, calcium saccharin is also soluble in water and used in food industry (WHO, 2008). In 1921 it was investigated by Paul that it is 300 times sweeter than ordinary sugar (sucrose) and its sweetness is greater in aqueous solution reported by Dubois in 1991(Paul, 1921). It is most widely used in beverages, food, cosmetics and pharmaceutical industries, as animal feed sweetener and electroplating of nickel (Salant, 1972). Sodium and Calcium ions are substituted with SO residue.(PanchalIshan I, 2014)

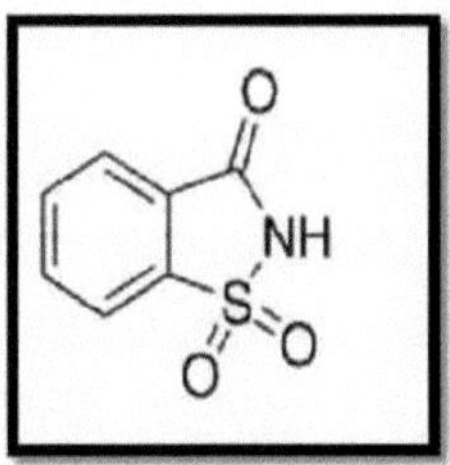

ASPARTAME

- **What is aspartame?**

Aspartame is sallow fragrance-free powder composed of aspartic acid and phenylalanine and methanol having molecular mass of $C_{14}H_{18}N_{2O}$. It was first approved by FDA in 1981 for its use in dry foods, in 1983 its use was prosper to carbonated drinks and in 1996 it was approved for general uses now its use is increased up to 4500 tons only in diet soda annually (Stonehouse *et al.,* 2013). As it is a dipeptide, its solubility is least in water and alcohol while more in fats and oils. Diets that contain aspartame are breath mint, cereals, chewing gums, ice creams, sugar free gelatin, jams, jellies and other soft drinks (Meister, 2006).

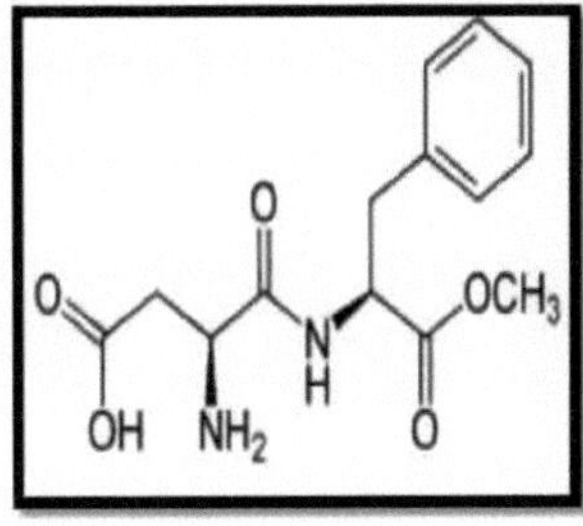

➢ What is the history of aspartame?

Aspartame was discovered as a novel sweetener in 1965. It was first authorized to enter the market in the United States in 1974. This authorization was suspended a few months later on the grounds that the first studies had not properly evaluated if aspartame could be toxic to the brain or cause brain cancer. A new assessment of those studies and the Examination of new data, led to a marketing authorization for solid food in 1981 and for Soft drinks in 1983. Aspartame was finally authorized as general sweetener in 1996. Up to Now, the safety of aspartame has been assessed by a Number of National and International Organizations.

An Acceptable Daily Intake (ADI) of aspartame for humans has been set at 40 mg/kg body weight per day by an international committee of experts. A debate on the risks to human health posed by the consumption of aspartame was relaunched, notably on the Internet, following an article published in 1996 which suggested there was a link between brain tumours and aspartame. Allegations claim that aspartame is responsible for a large number of adverse health effects such as multiple sclerosis, lupuserythematous, Gulf War Syndrome, brain tumours, epileptic seizures, complications of diabetes, etc. Aspartame is approved by the help of Dr.James Schlatter a medical researcher by Searle pharmaceutical. In 1965.

➢ What are the uses and properties of aspartame?

Aspartame is marketed as table sweetener (for example, Canderel [see Annex 1, p. 6] ®,NutraSweet [see Annex 3, p. 7] ® It is also incorporated in a number of foods stuffs Throughout the world, including drinks, desserts and sweets (European code E951). It is a white, odourless powder, approximately 200 times sweeter than sugar, manufactured by combining phenylalanine and aspartic acid. Its main impurity is diketopiperazine that has no sweetening properties. Aspartame is stable in the dry state and in frozen products. However, when stored in liquids at more than 30°C, it progressively converts into diketopiperazine, which is partially degraded into methanol, aspartic acid and phenylalanine. These transformations result in the loss of sweetness. Therefore, aspartame cannot be used in cooked or sterilized foods.

➢ How much aspartame do people consume?

The Acceptable Daily Intake(ADI) of 40 mg/kg body weight per day set by the committee of experts of the Food and Agriculture Organization (FAO) and the World Health Organization (WHO) is not likely to be exceeded, even by children and diabetics. A European Commission (EC) report gives a theoretical maximum estimate for adults' consumption of 21.3 mg/kg body weight per day of aspartame. However, the actual consumption is likely to be lower, even for high consumers of aspartame. The report also gives refined estimates for children which show that they consume 1 to 40% of the Acceptable Daily Intake. Other reports in Europe use actual food consumption data and actual sweetener levels in foods to estimate that high level intakes for the general population vary between 2.8 and 7.5 mg/kg body weight per day. People with diabetes are high consumers of foods containing aspartame; their highest reported intake varies between 7.8 and 10.1 mg/kg body weight per day.

➢ What happens to aspartame once it is ingested?

Following ingestion, aspartame is broken down in the digestive tract to form aspartic acid, phenylalanine and methanol. Therefore, hardly any aspartame gets in the blood. The body rapidly metabolises both aspartic acid and methanol, without significantly increasing their concentration in the bloodstream, even for aspartame taken as a single dose equivalent to the entire Acceptable Daily Intake (ADI). At high doses usually surpassing the ADI, the level of phenylalanine in blood may increase with the dose of aspartame given. However, high doses generally do not increase the blood level more than a normal meal - except for individuals affected by phenyl ketonuria disease (homozygous PKU).

SOME SYMPTOMS OF

SUCRALOSE

SIDE EFFECTS

1.

DIGESTION

Studies have shown that artificial sweeteners can wreak major havoc with your digestion, causing diarrhea, gas, and bloating.

2.

DISRUPTS YOUR INTESTINAL FLORA

One animal study found that consuming sucralose decreased the good bacteria in your digestive tract while increasing bad bacteria in your stool.

3.

PREVENTS ABSORPTION OF MEDICATION

Consuming sucralose may inhibit the proper absorption of certain medications, decreasing their potency. These drugs include medications for heart disease and cancer.

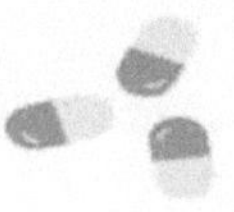

4.

HEADACHES

A study released in 2008 that artificial sweeteners may be responsible for triggering nagging headaches. This artificial sweetener lineup includes sucralose.

5.

ALLERGIES

Some people have reported that consuming sucralose has led to a host of symptoms resembling an allergic reaction.

In laboratory studies:

- Aspartame did not induce genet9ic mutations.
- A study on mice showed no cancer effects. A first study on rats fed with very high doses of aspartame (1000 to 6000 mg/kg body weight per day) found a higher incidence of brain tumours. This study was contradicted by two subsequent ones. Therefore, it was concluded that aspartame does not cause brain cancer in laboratory animals.

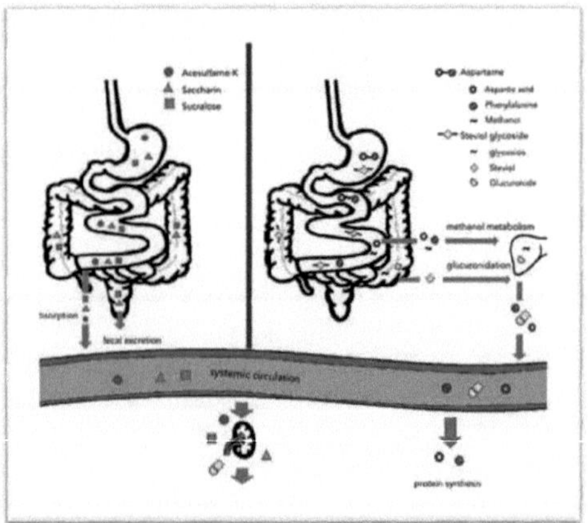

- Diketopiperazine does not cause cancer or genetic mutation in rats or mice. *4.1.2* In 1996, Olney published an article linking the consumption of aspartame and the occurrence of brain tumours in the United States, which has been criticized by a number of scientists. Subsequent studies did not find such a link. In France, the sale of aspartame did not increase the frequency of brain tumours.

➢ Aspartame affect reproduction and development

In laboratory animal tests, no effects on reproduction and development was observed below 4.000 mg/kg body weight per day. At higher doses, some pups grew marginally slower and weighed slightly less than normal because they ate less. No other effects of aspartame and its breakdown products were observed on reproduction and development (including neurodevelopment).

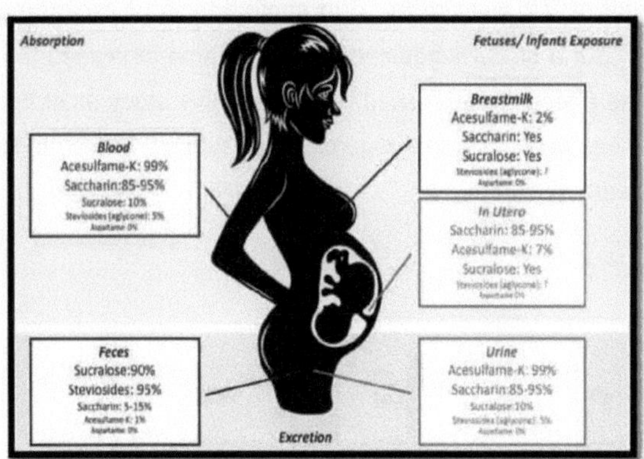

➢ Can aspartame produce neurological disorders?

Following the marketing of aspartame, some consumers complained of neurological or behavioural effects. These complaints received special consideration because some experiments in rats had shown that very high doses of aspartame (1000 mg/kg body weight per day) could alter the concentrations of some important substances (neurotransmitters) in the central nervous system. However, it appears that these effects on neurotransmitters are not consistent and could not be reproduced in later studies. About 10% of aspartame breaks down into methanol, which is known to be toxic. However, toxic effects on vision and the central nervous system only occur at doses of methanol 100 times higher than could be produced from the amount of aspartame in one litre of "diet" soft drink.

➢ Can aspartame affect behavior, cognition or mood?

Some years ago, it was suggested that aspartame could have effects on human behavior and cognition. However, studies on laboratory animals showed no adverse effects on behaviour and cognition, even at very high doses (up to 2000 mg/kg body weight per day). Also, studies in healthy adults and children and in people heterozygous for phenylketonuria (PKU) disease failed to show effects of aspartame on behaviour, mood or learning. But a study suggested that aspartame increased the frequency and severity of adverse effects in depressed individuals; this study must however be taken with caution as there were criticisms regarding the authors' interpretation and because too few subjects were evaluated.

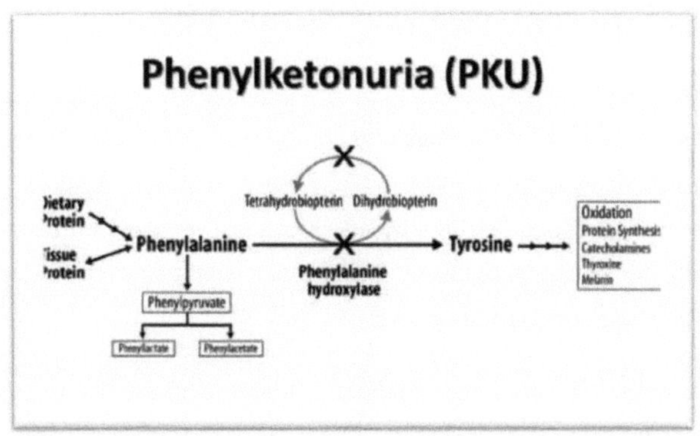

➢ Does aspartame cause headaches?

Headaches are one of the most commonly reported symptoms. Three studies on humans indicate a possible association between aspartame intake and headaches. However, it is not possible to draw conclusions. A more recent study in a controlled environment on individuals complaining of aspartame-related headaches concluded that aspartame was no more likely to trigger headaches than placebo

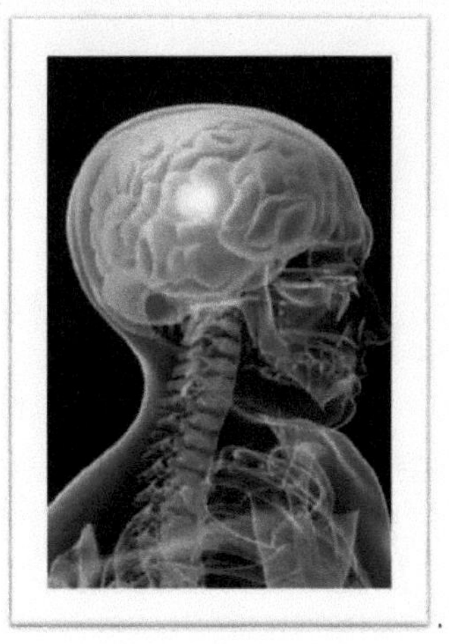

➢ Does aspartame trigger epileptic seizures?

Some websites report testimonies of people who identify aspartame as the cause of their health problems and epileptic seizures. A few studies have linked the consumption of large amounts of aspartame and the triggering of epileptic seizures. They suggest that aspartame could cause seizures by affecting the synthesis of neurotransmitters in the brain. Also, some animal studies indicate that aspartame reduces the threshold of sensitivity to chemically induced seizures.

Another study reported that aspartame could increase the duration of certain types of epileptic seizure in children. Effects of phenylalanine, aspartic acid and methanol on seizures have been reported, but under unusual conditions, such as high doses, particular sensitivities or rare types of seizures. This relationship has been refuted by a large number of scientists, who base their opinions on many animals and humans studies. The Epilepsy Institute in the USA has also concluded that aspartame is not the cause of epileptic seizures.

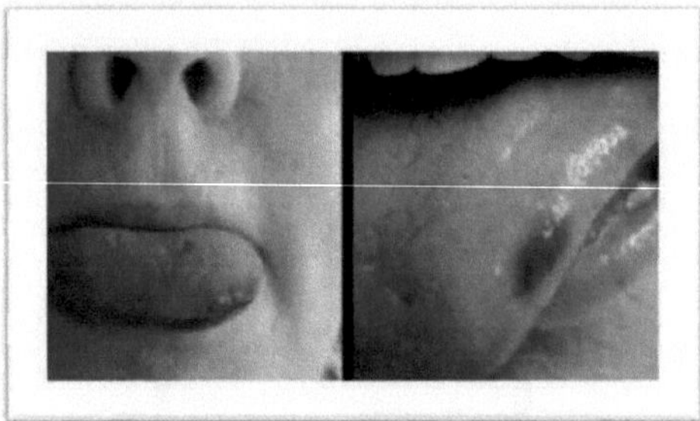

➢ Aspartame cause other health effects:

Some individuals have attributed their allergic reactions to aspartame. However, two studies on such individuals indicated that their allergies were no more likely to be caused by aspartame than by placebo. Other studies focusing on the effects of aspartame on hunger and food intake, reported that aspartame did not cause an increase in calorie consumption or body weight.

ACESULFAME POTASSIUM (ACE-K)

Chemically Ace-K is crystalline powder made of 5,6-dimethyl-1,2,3-oxathiazin-4(3H)-one-2,2-dioxide, first discovered in 1967 by Clauss and Jensen. The sweetness of is dependent upon its ring structure any modification in its structure results in decrease in sweetness.

Acesulfame is highly acidic compound the neutrality is achieved by potassium hydroxide (Arpe, 1978). Sweetness of Ace-K is 200 times greater than that of sucrose, but least sweet than that of saccharin and cyclamate and can be preserved for six year or more even at room temperature. The time of sweetness is slower and doesn't continue for longer time, but at high concentration it may taste bitter(Nabors, 2012).It is abundantly used in baked food as well as in soda bottles due to its stability in heat, aqueous environment and long half-life (Walters, 2009; Nabors, 2012).

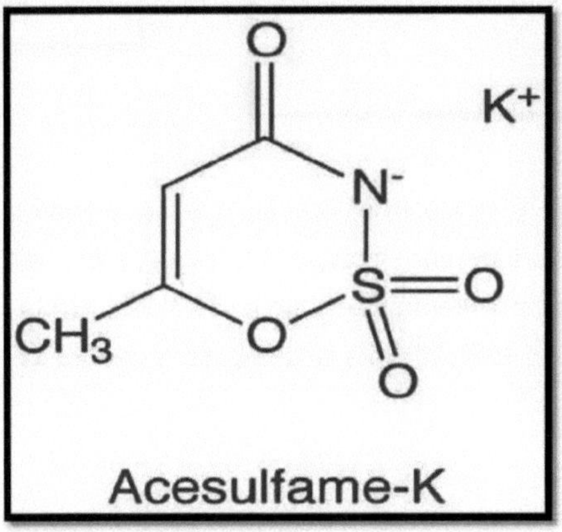

Acesulfame-K

NEOTAME (N-(N-(3,3-DIMETHYLBUTYL)-L-A-ASPARTYL)-L-PHENYLALANINE 1-METHYL ESTER)

Neotame is derivative of phenylalanine and aspartic acid having sweeting influence 7000-8000 times greater than that of table sugar.As it is structurally resembled with aspartame its metabolism is same as that of aspartame, but it is highly stable in powdered form(Panchal *et al.,* 2014; Nabors, 2012). Neotame is made powder by de-esterification at low temperature and are high functional stability under range of temperature and pH.Neotame doesn't undergo intra-molecular cyclization so no DKP formed after its degradation (Nabors, 2012).

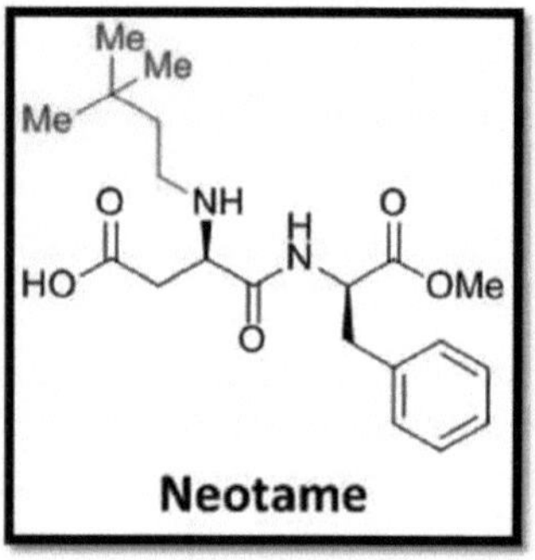

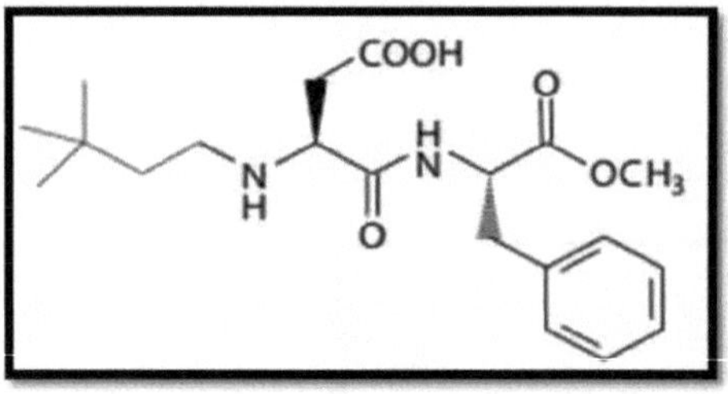

Neotame

Neotame is 30-60 time sweeter than aspartame due to modification in aspartate residueby tertiary butyl group. This modification helps in de-esterification resulting in easy clearance from plasma within 72 hours (EFSA, 2007). It's long term exposure may leads to low birth rate and excessive weight loss (Whitehouse *et al.*, 2008).

SUCRALOSE

Sucralose (also known as Splenda®) was approved by the FDA as a tabletop sweetener in 1998, followed by approval as a general purpose sweetener in 1999. Sucralose is made from sucrose by a multistep patented manufacturing process that selectivelyreplaces three hydroxyl with chlorine atoms. This molecular change makes sucralose 600 timessweeter than sugar.

It is used in beverages, frozen desserts, chewing gum, baked goods and other foods. This cannot be construed as natural, sucralose contains chlorine. It is heat stable, meaning that it does not break down when cooked or baked.

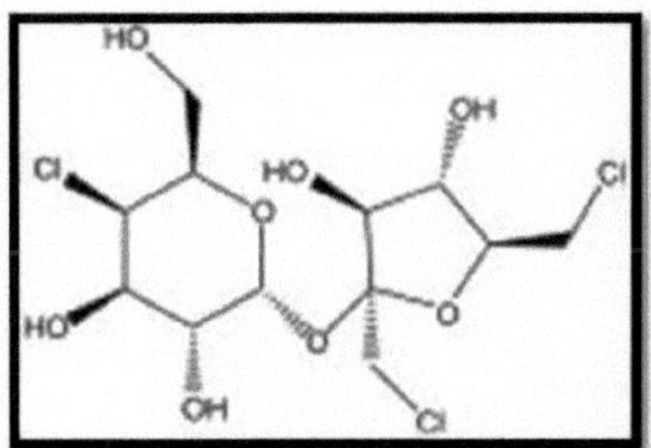

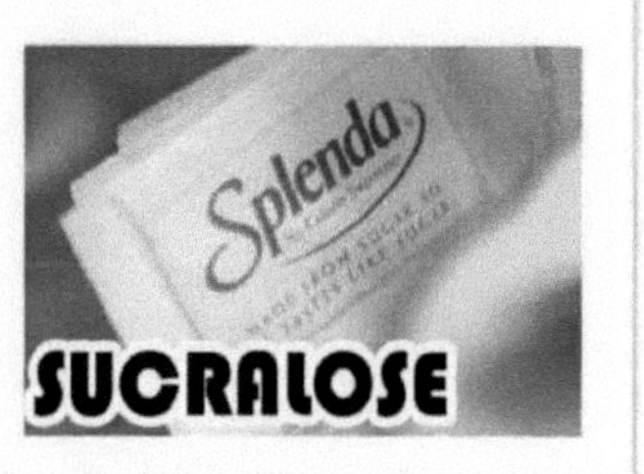

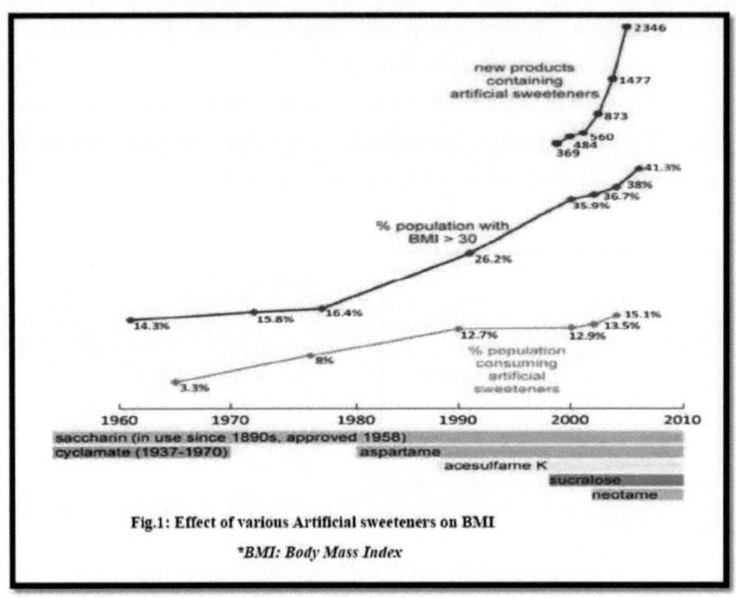

Fig.1: Effect of various Artificial sweeteners on BMI

**BMI: Body Mass Index*

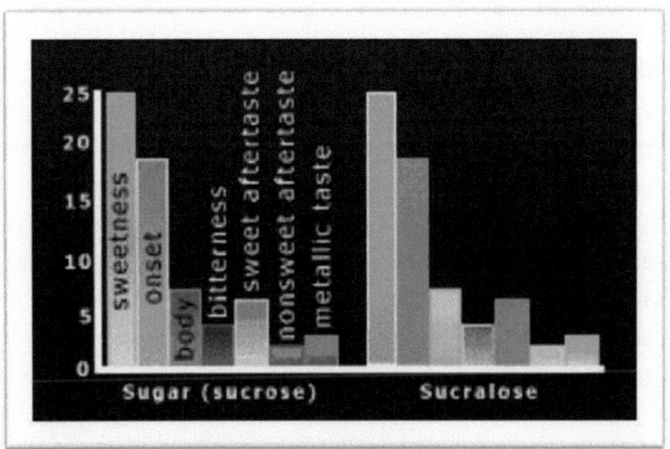

VITRO AND IN VIVO RODENTS EXPERIMENT:

Although the majority of in vitro and in vivo studies in rodents detected significant effects of sucralose on physiological processes involved in nutrient absorption, findings of similar effects in humans have been inconsistent. Sucralose delivered in isolation (e.g., without co-administration of glucose) to human volunteers by oral (Ford et al., 2011) or by intragastric routes (Ma et al., 2009; Steinert et al., 2011) exerted no marked effect on GLP-1. Oral consumption of sucralose without co-administration of glucose (Brown et al., 2011) produced no significant effect on blood glucose levels. Sucralose delivered by intraduodenal infusion in combination with glucose also exerted no marked effect on blood glucose or plasma GLP-1 (Ma et al., 2010). However, GLP-1 was elevated in human subjects (both in healthy volunteers and in individuals with type 1 diabetes) who drank a caffeine-free diet soda sweetened with sucralose and ace-K when compared with a carbonated water control; in both conditions subjects consumed the test solution (diet soda or carbonated water) 10 min prior to a glucose load (Brown et al., 2009, 2012). Under similar test conditions, obese women displayed elevated glucose and insulin levels when exposed to sucralose alone

Effect of Sucralose on Body Weight in Humans:

The effect of chronic oral consumption of sucralose on body weight at levels approved by the FDA and EU has not been studied prospectively in adults. In a lab setting, ingestion of sucralose tends to increase intake after a 60-min interval (Anderson et al., 2002); however, it is not yet known whether this finding transfers into chronic body weight gain in free-living adult populations. In an 18-mo trial with children, participants were randomly assigned to receive an 8-oz can per day of either a noncalorically sweetened or a sugar-sweetened beverage that provided 104 kcal (de Ruyter et al., 2012). Each can of sugar-free beverage contained 34 mg sucralose and 12 mg acesulfame-K, and the mean sucralose dosage for the sugar-free group was 1.1 mg/kg/d. The mean duration of the study was 541 d (77.3 wk), during which 477 of 641 children completed the intervention by consuming an average of 5.8 beverages per week. Measurement of urinary sucralose levels was one of several markers used to monitor compliance with the protocol. The calorie consumption from these beverages was 46,627 kcal greater for children in the sugar-sweetened group than in the sucralose-sweetened group (5.8 × 77.3 × 104). In spite of this highly significant difference in calories consumed from the beverages, the total weight gain over this 18-mo study was only 1 kg greater for children in the sugar-sweetened group compared to sucralose group.

No explanation was provided to account for the small difference in weight gain given the large difference in caloric consumption from the beverages. However, one possible explanation is that the children who consumed the sugar-sweetened beverages compensated by reducing their food intake. A control group that ingested water as a comparison was not included. Another study in adolescents showed no consistent reduction of weight gain at a 2-year follow-up when their families were supplied with artificially sweetened beverages in order to reduce their consumption of sugar-sweetened sodas (Ebbeling et al., 2012).

Effect of Sucralose on Body Weight in Animals:

In rats, a sucralose dosage of 1.1 mg/kg/d (equivalent to that provided to children by de Ruyter et al., 2012) produced a significant body weight gain (10.9% greater than controls) over a 12-wk period (Abou-Donia et al., 2008). However, weight gain relative to control was not observed in rats at higher dosages of 3.3, 5.5, and 11 mg/kg/d sucralose. After a 12-wk recovery period from sucralose treatment, there were significant increases in body weight for rats previously treated with sucralose relative to control animals for 2 of 4 sucralose dosage groups. After recovery from 1.1 mg/kg/d, there was a 17.1% rise in body weight relative to control animals, and after recovery from 5.5 mg/kg/d, a 21.3% elevation. In the human nutrition literature, a weight difference ≥5% is considered clinically significant (Wengreen and Moncur, 2009). Two possible explanations (described in more detail later in this review) for the significant body weight gain at the lowest (not highest) dosage of sucralose by the end of treatment period include: (1) Sucralose may no longer be bioavailable as an intact molecule at higher dosages due to increased intestinal expression of (and metabolism by) CYP enzymes, and (2) intestinal bacteria that normally reclaim calories from undigested dietary nutrients in the intestines are suppressed to a greater degree at the higher dosages. The finding of a weight gain after a 12-wk recovery from sucralose may be due to a partial rebound in the number of bacteria that ferment complex dietary carbohydrates to form short-chain fatty acids that are subsequently absorbed.

Animal data regarding body weight gain reported in historical subchronic and chronic toxicity studies utilized sucralose dosages that far exceed levels approved for use in foods by the FDA (U.S. FDA, 1998) and EU (2004), and the results on body weight gain in these toxicity studies are inconsistent and conflicting. Increases in mean body weight gain were found in sucralose-treated beagle dogs relative to controls at 0.3 to 3% of the diet (approximately 90 to 900 mg/kg/d respectively), and this weight gain was accompanied by a rise in food consumption (Goldsmith, 2000b). Body weight gain in rats, however, appeared to be dependent in part on the route of delivery of sucralose. Rats that consumed high doses of sucralose in toxicity tests by the oral route (dietary administration) showed significant decreases in body weight gain relative to controls in spite of relatively small reductions in food consumption. These disproportionately large declines in body weight gain despite only small decreases in intake were attributed to low palatability of sucralose at high concentrations. Body weight changes in rats that received high dosages of sucralose by gavage (which circumvents taste issues associated with dietary administration of an unpalatable test material) were variable with losses, gains and no marked changes reported (Goldsmith, 2000b; U.S. FDA, 1998). These animal feeding results are not directly transferable to humans, however, because the dosages delivered to animals in these toxicity tests were significantly higher than usage levels approved for humans.

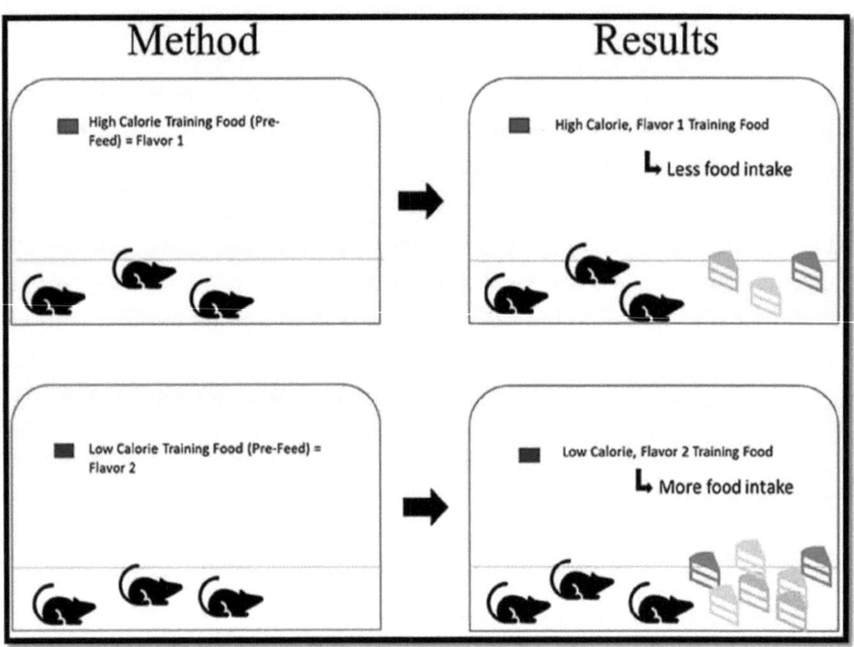

Learned Associations and High-Potency Sweeteners as a Group:

A learning paradigm has been proposed to explain how high-potency sweeteners as a group may influence body weight regulation and glucose homeostasis (Swithers et al., 2010, 2012; Swithers, 2013). This paradigm is based on the principle that humans and non-human animals learn the relationships between the sensory properties of foods (e.g., taste, smell, texture) and their postingestive nutritive consequences to maintain energy balance (Warwick and Schiffman, 1991; Swithers and Davidson, 2008; Swithers et al., 2009, 2010). Throughout evolutionary history (at least until the recent introduction of high-potency sweeteners), learned sweet taste cues have been reliable predictors of the energy density of food. However, the introduction of high-potency sweeteners that have negligible utilizable calories into the food supply reduced the validity of sweet taste as a signal for calories. Uncoupling the relationship between sensory properties of foods and their caloric content was shown to contribute to weight gain in rats (Warwick and Schiffman, 1991). Using learning paradigms with a rat model, Swithers and colleagues (Swithers and Davidson, 2008; Swithers et al., 2009, 2010) found that consumption of foods and fluids containing high-potency sweeteners (saccharin, acesulfame-K, and/or steviol glycosides were used in these experiments) interfered with the ability of sweet taste to predict caloric consequences and thus disrupted energy regulation. Further, the nonpredictive sweet–calorie relationship was persistent and difficult to reverse even after predictive sweet–calorie training with glucose (Swithers et al., 2009). Thus, exposing rats to high-potency sweeteners degraded the consistent and predictive relationship between taste, energy intake, and body weight regulation.

The degree to which these animal-learning studies of sweeteners can be extrapolated to humans and to sucralose in particular has not yet been determined. Humans can distinguish among many different sweetener types by taste (Schiffman et al., 1979, 1981, 1995; Schiffman and Gatlin, 1993), and functional magnetic resonance (fMRI) imaging studies also indicate that the human brain signals distinguish caloric sweeteners (e.g., sucrose) from noncaloric sweeteners (e.g., sucralose and saccharin)

(Frank et al., 2008; Haase et al., 2008). However, neuroimaging studies by Rudenga and Small (2012) revealed that routine use of high-potency sweeteners as a group altered brain responses to sucrose in the amygdala and insula as determined by MRI scanning. Data suggested that these alterations in brain activity were due to degradation or uncoupling of the predictive relationship between sweet taste and its postingestive consequences, as reported previously in rat models (Swithers et al. 2010). The findings of Rudenga and Small (2012) are consistent with those of Green and Murphy (2012), who reported that higher order reward regions of the brain are activated to a greater extent by sweeteners in young adult diet-soda drinkers compared to nondrinkers. These brain imaging studies in conjunction with epidemiological data that associate artificial sweetener use with weight gain suggest that high-potency sweeteners interfere with learned sweet–calorie relationships in humans as well as animals (Swithers, 2013).

Effect of Sucralose in Patients With Diabetes:

A study of patients with diabetes (Grotz et al., 2003) reported no significant effect of sucralose on glycosylated hemoglobin (HbA1c) as a marker of the average plasma glucose concentrations over approximately 3 mo. Grotz et al. (2003) instructed patients with diabetes to self-administer capsules of sucralose twice daily over a 3-mo period; however, data on compliance with self-administration (e.g., urinary sucralose levels) and body weight changes were not reported. Daily supervision of capsule administration, rather than self-administration, is scientifically prudent for studies of patients with diabetes because noncompliance with treatment regimens is reportedly high in this population (WHO, 2003). Further, no apparent information of dissolution characteristics of the capsule was provided; thus, it is unknown where sucralose was released in the GIT. Future studies of sucralose on diabetes management will require randomized trials that are carefully supervised and control for the variables involved.

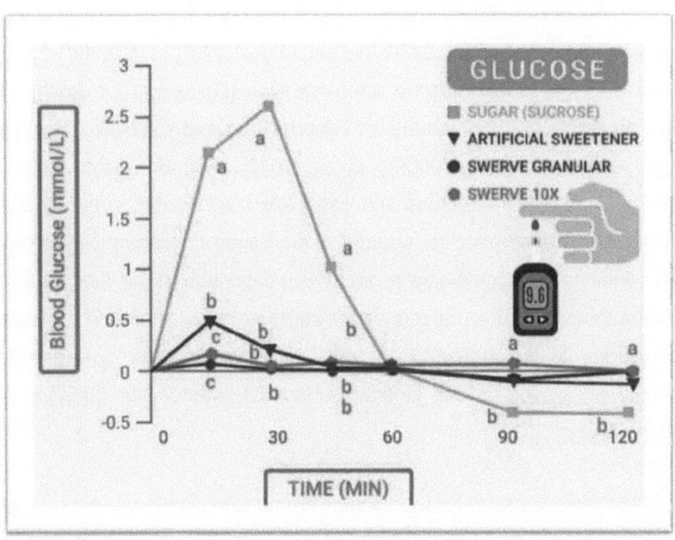

EFFECTS OF SUCRALOSE ON PRESYSTEMIC DETOXIFICATION MECHANISMS AND IMPACT ON BIOAVAILABILITY OF THERAPEUTIC DRUGS

Background: Role of Intestinal Transporters and CYP Metabolizing Enzymes in Bioavailability of Xenobiotics:

Orally administered xenobiotics such as therapeutic drugs and artificial food additives reach the small intestine, where they cross the intestinal membranes to be transported into the hepatic portal system and ultimately to the systemic circulation. During their transit from the GIT to the systemic circulation, xenobiotics interact with metabolic enzymes and transporters in the GIT and liver that limit their systemic bioavailability. The decrease in the concentration of a xenobiotic compound as it passes through the GIT and liver is termed the "first-pass effect" (Riegelman and Rowland, 1973; Iwamoto and Klaassen, 1977; Shimomura et al., 2002; Watkins, 1997; Paine and Thummel, 2003; Paine, 2009).

The intestinal component of the first-pass effect is especially significant in the disposition of sucralose because the majority of this OC sweetener is reportedly unabsorbed from the small intestine after ingestion (Grice and Goldmith, 2000) even though it is an amphiphilic compound with appreciable lipid solubility (Hough and Khan, 1978; 1989; Anderson et al., 2006; Li et al., 2010). Further, higher doses of orally administered sucralose in humans are associated with lower urinary excretion that suggests sucralose absorption is reduced at increased concentrations (Roberts et al., 2000). The findings by Abou-Donia et al. (2008) described in the next section suggest that the efflux transporter P-gp and metabolism by intestinal CYP limit oral absorption of sucralose (and/or its metabolites) from the GIT. P-gp is an energy-dependent "pump" that is highly expressed at the luminal surface of enterocytes (Lin and Yamazaki, 2003; Marchetti et al., 2007).

It is a member of the ATP-binding cassette (ABC) transporter superfamily and a product of the *ABCB1* gene (also known as *MDR1*). P-gp serves as a barrier to harmful chemicals (as well as many therapeutic drugs) by transporting them out of enterocytes, back into the intestinal lumen (Sparreboom et al., 1997; Masuda et al., 2000; Westphal et al., 2000; Suzuki and Sugiyama, 2000; Drescher et al., 2003; Abu-Qare et al., 2003; Fang et al., 2009). P-gp interacts with many structurally diverse hydrophobic and amphiphilic compounds (Garrigues et al., 2002; Marchetti et al., 2007; Yang et al., 2009) including OC drugs (Polli et al., 2001; Boulton et al., 2002; Wang et al., 2001; 2002; 2008) and OC pesticides (Bain and LeBlanc, 1996).

Intestinal metabolism by members of the CYP superfamily of heme-containing enzymes contributes significantly to the first-pass effect (Kolars et al., 1991; Watkins, 1992, 1997; Paine et al., 1996, 2006; Thummel et al., 1997; Lampen et al., 1998; Hall et al., 1999; Wacher et al., 2001; von Richter et al., 2001): The CYP superfamily is divided into families (including CYP1, CYP2, and CYP3, which are responsible for metabolism of drugs and xenobiotics) and into subfamilies labeled with a letter, such as CYP3A or CYP2D. CYP3A has a broad substrate specificity and is a predominant isoform of CYP expressed in the upper intestine (Watkins et al., 1987; Kolars et al., 1994; Guengerich,

1999; Zhang et al., 1999; Mitschke et al., 2008; Takara et al., 2003; Matsubara et al., 2004). Organochlorine drugs such as midazolam undergo significant intestinal metabolism by the CYP3A subfamily (Paine et al., 1996; Thummel et al., 1996; Higashikawa et al., 1999; Bruyère et al., 2009). The catalytic activities of intestinal and hepatic CYP3A are independently regulated so that measures of CYP-mediated metabolism of xenobiotics in the liver do not necessarily predict the effects of CYP-mediated metabolism in the intestine (Hakkak et al., 1993; Lown et al., 1994; Aiba et al., 2003; Paine and Thummel, 2003; van Herwaarden et al., 2009).

Although the roles of P-gp and CYP in limiting the bioavailability of xenobiotics are different, there is significant overlap in compounds that are effluxed by P-gp and those metabolized by CYP3A (Watkins, 1997; Wacher et al., 1995, 2001, Schuetz et al., 1996; Benet et al., 1999). While P-gp effluxes xenobiotics back into the intestinal lumen, CYP enzymes promote elimination of xenobiotics through oxidation and reduction reactions that render chemicals more polar and water-soluble. Studies of xenobiotic detoxification and clearance indicate that co-localization of P-gp and CYP3A within enterocytes facilitates an interactive process by which the bioavailability of compounds that are dual P-gp/CYP3A substrates is reduced. P-gp enhances presystemic metabolism in the intestine by increasing the duration of exposure to CYP3A via repeated efflux and reabsorption into enterocytes. This repetitive recycling process that engages both intestinal P-gp and CYP3A provides a barrier to absorption (Benet, 2009), and may account for the low systemic oral bioavailability of sucralose.

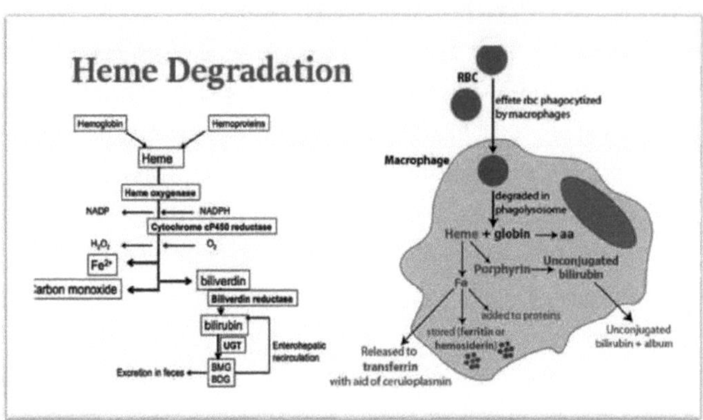

Splenda (Active Ingredient: Sucralose) Increases the Expression of Intestinal P-gp and Two CYP Isoforms:

Abou-Donia et al. (2008) reported that Splenda, a commercially available form of sucralose, increased the expression of intestinal P-gp and CYP in male Sprague-Dawley rats. Splenda contains the high-potency OC sweetener sucralose along with the fillers, maltodextrin and glucose. Splenda was administered for 12 wk by oral gavage at sucralose dosages equivalent to 0 (vehicle control), 1.1, 3.3, 5.5, or 11 mg/kg/d. All sucralose dosages tested fell below the ADI approved for use in the food supply by the European Union (EU, 2004). The two lower dosages fell below the ADI for sucralose approved by the FDA (1998). After 12 wk, half of the animals from each group were sacrificed to measure the expression of intestinal P-gp, CYP3A, and CYP2D in the jejunum and ileum. The other half of the animals were allowed to recover from treatment (no Splenda) for another 12 wk, after which expression of intestinal P-gp, CYP3A, and CYP2D was again measured. P-gp and CYP expression were assessed using antibodies that detected P-gp, CYP3A protein, and CYP2D1, the rat isozyme analogous to human CYP2D6 (Laurenzana et al., 1995). These antibodies were selected because the CYP3A and CYP2D subfamilies are collectively involved in the metabolism of more than 70% of medications (Dantzig et al., 1999; Felmlee et al., 2008; Ingelman-Sundberg, 2005).

METABOLIC FATE AND SAFETY PROFILE OF SUCRALOSE METABOLITES

Metabolites of sucralose have been detected in the feces and urine of rats and humans by thin-layer chromatographic (TLC) methods, but the chemical identities of these metabolites have not yet been established (Sims et al. 2000; Roberts et al. 2000). Thin-layer chromatograms (TLC) of methanolic fecal extracts following administration of ^{14}C-sucralose to rats (Sims et al. 2000) and humans (Roberts et al. 2000) suggested that sucralose is metabolized in the GIT. A comparison of TLC radiochromatographic profiles of methanolic fecal extracts following a single intravenous (iv) and a single oral administration of ^{14}C sucralose in two rats is shown in Figures 1a and 1b, respectively (from Sims et al., 2000). The profile in Figure 1b (the peaks are enlarged to the right) from the rat that received oral ^{14}C-sucralose yielded a broader trace that contained multiple, closely eluting peaks of approximately equal height when compared to the thinner profile with one peak from the rat that received the iv dose. The multiple peaks in the trace in Figure 1b indicate the presence of at least two radioactive chemicals in the fecal material; that is, sucralose underwent metabolism and was not excreted unchanged in the feces. In addition, the R_f values (i.e., relative distance) of the peaks in Figure 1b from the oral dose are shifted somewhat to the left of the R_f value of the putative sucralose peak from the IV dose in Figure 1a. This suggests that at least one of the peaks in Figure 1b represents a chemical other than sucralose itself. The rat that received the oral dose had been maintained on a sucralose-containing diet for 85 wk that, according to the findings of Abou-Donia et al. (2008), would most likely have enhanced the expression of CYP. TLC radiochromatographic profiles of fecal extracts from a single human subject using two different solvents also indicate the presence of several peaks (Roberts et al., 2000).

Although TLC is a straightforward technique to separate component compounds in mixtures, it cannot be used to identify specific metabolites. Use of liquid chromatography–mass spectrometry (LC-MS) enables the chemical identification of sucralose metabolites. After the metabolites have been systematically identified, they can be evaluated for safety. Metabolites of drugs with exposures that are >10% of the administered dose or systemic exposure are recommended for safety assessment by regulatory agencies (Robison and Jacobs, 2009).

EFFECT OF SUCRALOSE ON THE NUMBER AND RELATIVE PROPORTIONS OF DIFFERENT INTESTINAL BACTERIAL TYPES

Studies of bacteria in culture media suggest that sucralose is not utilized as a growth substrate by microorganisms from the oral cavity (Young and Bowen, 1990) or from soil (Lappin-Scott et al., 1987; Labare and Alexander, 1993, 1994). These findings raise the question of whether the presence of unabsorbed sucralose or its metabolites in the GIT affect the metabolic activity and composition of GIT microflora of humans and non-human animals. Gut microflora perform many useful functions, including fermentation of complex dietary carbohydrates with concomitant formation of short-chain fatty acids (SCFA), synthesis of vitamins (e.g., B and K), modulation of immune responses, regulation of postnatal gut development, inhibition of pathogens, absorption of calcium and magnesium, and metabolism of drugs (Albert et al., 1980; Cummings and Macfarlane, 1991, 1997; Shearer, 1995; Hill, 1997; Bauer, 1998; Holzapfel et al., 1998; Chonan et al., 2001; Topping and Clifton, 2001; Hart et al., 2002; Teitelbaum and Walker, 2002; Fooks and Gibson, 2002; Guarner and Malagelada, 2003). Intestinal microbiota may modulate the expression of CYP, conjugating enzymes including UGT, and P-gp (Nicholson et al., 2005; Ueyama et al., 2005; Jia et al., 2008; Claus et al., 2008; Björkholm et al., 2009; Meinl et al., 2009), which were reported to play a role in the disposition of sucralose. Intestinal bacterial composition was also shown to play a role in obesity (Duncan et al., 2007; Ley et al., 2006; Turnbaugh et al., 2006; Turnbaugh and Gordon, 2009).

Sucralose Administered as Splenda Reduces Bacterial Counts in the Gastrointestinal Tract and Alters Their Relative Proportions

Abou-Donia et al. (2008) found that sucralose delivered as Splenda reduced the numbers and altered the composition of microbiota in the GIT of male

Sprague-Dawley rats. Fecal samples were collected each week during the 12 wk of treatment and for 12 wk of recovery from treatment for bacterial culture studies, quantification of fecal pH, and histopathological studies of the colon. Data showed that bacterial counts in the GIT from daily sucralose ingestion decreased progressively and monotonically in a methodical pattern during each successive week of sucralose treatment. Table 3 shows the percent difference in bacterial counts for rats treated with sucralose relative to counts from control rats at the end of the 12-wk treatment period and at the end of the 12-wk recovery. The numbers of total anaerobes, bifidobacteria, lactobacilli, *Bacteroides*, clostridia, and total aerobic bacteria were significantly decreased at the end of the 12-wk treatment period with losses up to 79.7% for lactobacilli; there was no significant treatment-related effect on enterobacteria. These changes in bacterial counts were accompanied by intermittent incidences of unformed or soft feces as well as histopathological changes in the colon, including lymphocytic infiltrates into epithelium, epithelial scarring, mild depletion of goblet cells, and glandular disorganization.

At the end of the 12-wk recovery period, the total anaerobes and bifidobacteria were still significantly lowered. Fecal pH increased during the treatment period (up to 7.4%) and remained significantly elevated at the end of the 12-wk recovery period. Abou-Donia et al. (2008) concluded that sucralose (administered as Splenda) reduced the number of indigenous intestinal bacteria, with significantly greater suppression for the generally beneficial anaerobes (e.g., lactobacilli, and bifidobacteria), and with less inhibition for more detrimental bacteria (e.g., enterobacteria). Further, the numbers of total anaerobes remained partially suppressed after a 3-mo recovery period.

ADVANTAME(N-[N-(3-(3-HYDROXY-4-METHYLOXYPHENYL) PROPYL]A-ASPARTYL]-L-PHENYLALANINE 1 METHYL ESTER

Advantame is an intensive sweetener derived from aspartame and vanillin. It is similar in structure to neotame except in side chain to the N- terminal of L-aspartic acid. Advantame is made through fermentation step and synthetic chemical step similar to aspartame (dipeptide formation) and the addition to the N-terminal of

L-aspartic acid the side chain (HMPA) 3-(3-hydroxy-4-methyloxyphenyl) propyonaldehyde. Advantame is about 20,000 times sweeter than sucrose and 100 times sweeter than aspartame and FDA has approved advantame as sweetener for general use in foods and beverages. Its clean sweet taste is very similar to aspartame with only a slightly longer sweetness.

Its heat stablility makes it suitable as a sugar substitute in baked goods and is less stable in the acidic conditions such as beverages products. Acceptable daily intake (ADI) for advantameis 5 mg/kg. body weight. Advantame can be used to partially replace sugar, high fructose corn syrup or other high potency sweeteners to reduce cost, calories, while maintaining the same taste profile of the products.

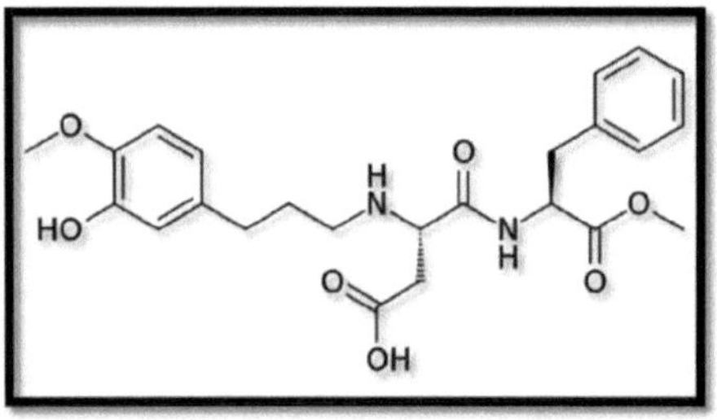

OTHER ARTIFICIAL SWEETNERS

Here are brief review of some other artificial sweeteners used as sugar substitutes

CYCLAMATE (SODIUM N CYCLOHEXILESULFAMATE)

Cyclamate or mostly in the form of its sodium salt Sodium cyclamate is an artificial sweetener. It is 30–60 times sweeter than sugar. FDA banned the use of cyclamate in 1969 after some studies in rats which claimed its risk of causing bladder cancer in humans. But after further studies regarding its carcinogenicity the researchers concluded that it cyclamate is not carcinogenic. Petition for the reapproval of cyclamate was field with the FDA but still in vein. However it is still used as artificial sweetener in many countries other than US. Mostly it is used with the other sweeteners especially its mixture having a ratio of saccharine- cyclamate (1:10) is common. It is less expensive than most sweeteners, including sucralose, and is stable under heating but with an unpleasant aftertaste. Cyclamate is used in cola, milk, Ice cream, soft drinks, canned food, biscuits, sweets, preserved fruits, pickles, wine, family seasoning, cooking. It is also used in pharmaceutical & cosmetic such as toothpastes, mouth freshness & lipstick.It is suitable use for diabetes, and high blood pressure people.

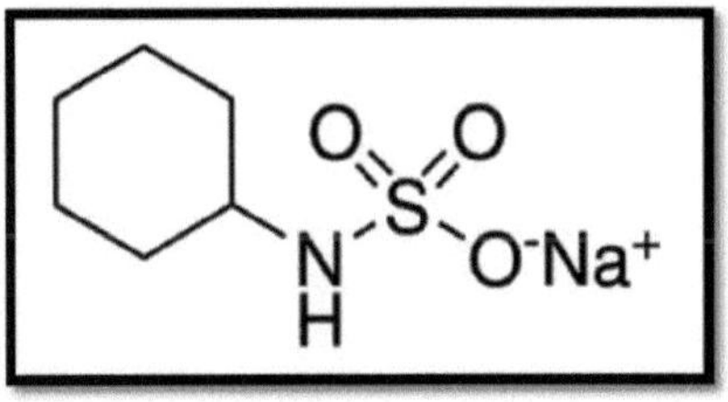

DULCIN

Dulcin is an artificial sweetener about 250 times sweeter than sugar discovered in 1884 by Joseph Berlinerbau. It was first mass-produced about seven years later. Despite the fact that it was discovered only five years after saccharin, it never enjoyed the latter compound's market success. Still, it was an important sweetener of the early 20th century and had an advantage over saccharin in that it did not possess a bitter aftertaste.

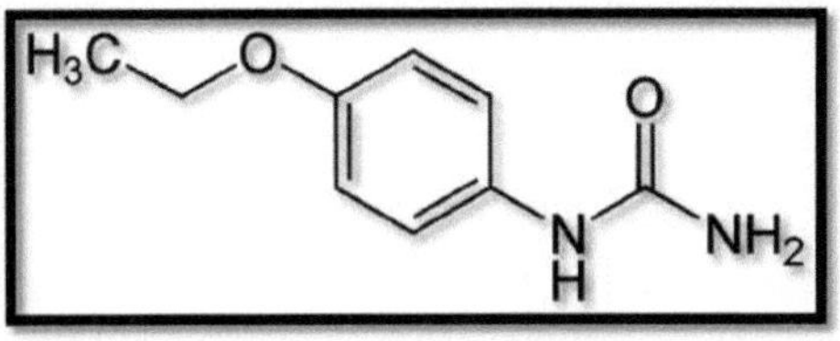

ALITAME [L-ALPHA –ASPERTYL-N-DALANINAMIDE]

Alitame is also a non-caloric artificial sweetener and about 200-300 times sweeter than sucrose and 10 times sweeter than aspartame. Chemically it is a dipeptide of L-aspartic acid and D-alanine, with a terminal *N*-substituted by tetramethyl-thietanly-amine.Alitame has been approved under the brand name **Aclame**for use in a variety of food and beverage products in Australia, New Zealand, Mexico and China. However in USA its petition as a sweetening agent or flavoring in foods has been withdrawn due to manufacturing cost. Acceptable daily intake (ADI) for alitame is 1mg/ kg. body weight. Alitame has a clean sweet taste with excellent stability at high temperature.

It is suitable for diabetics and also for weight loss as its caloric contribution to the diet is negligible. It is safe for teeth and for human consumption. It also show good effect when combined with other sweeteners.

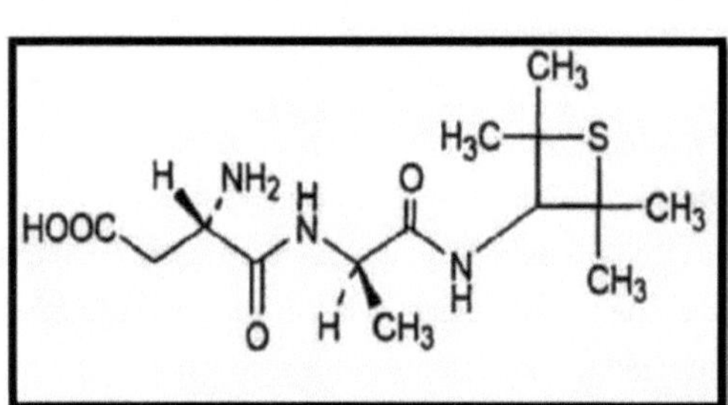

SUGAR ALCOHOLS AS SUGAR SUBSTITUTE

XYLITOL

Xylitol a sugar alcohol, present naturally in corn cobs, fibrous fruits and vegetables and some hardwood tress. It is produced naturally in our bodies as well. The production on commercial scale is achieved by the catalytic hydrogenation from the corresponding sugar. Xylitol is produced from xylose. Xylitol is used as a sugar substitutebecause as sweet like sucrose but only has two thirds of the energy. It is used in cooking, baking, in beverages, chewing gum, mints, and other products such as nasal and mouth washes. Xylitol is a five-carbon sugar and it possess the antibacterial properties. Xylitol prevents the tooth decay process as when it is consumed it starves the microorganisms which is helpful to demineralize the damage teeth and repair minor cavities. Xylitol does not contribute to high blood sugar levels or the resulting hyperglycaemia caused by insufficient insulin response.

It may also have potentialas a treatment for osteoporosis. Xylitol based chewing gum can help to prevent ear infectionsas the act of chewing and swallowing helps with the disposal of earwax and clearing the middle ear, while the xylitol prevents the growth of bacteria in the Eustachian tubes.

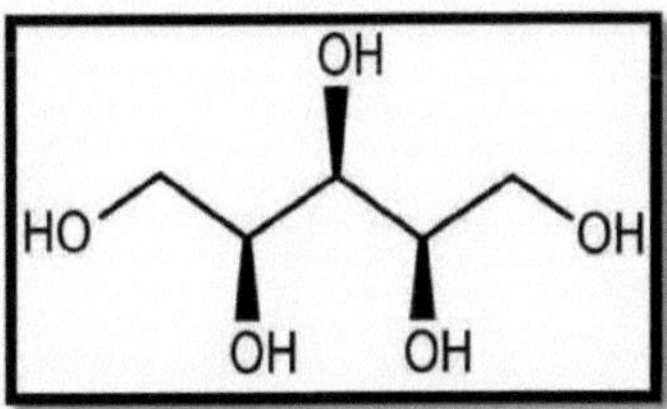

SORBITOL

Sorbitol is also a sugar alcohol and also known as glucitol is produced from glucose. It is slowly metabolized by the body. The main uses of sorbitol are in diet foods, sugar-free chewing gums, mints and also in cough syrups. It is also found naturally in the fruit of various trees of the genus Sorbus. It is beneficial for diadetic patients and also prevent tooth decay. However some adverse effects such as abdominal pain, gas and diarrhea was observed but when large amount consumed. Fructose malabsorption and bowel syndrome increases due to sorbitol.

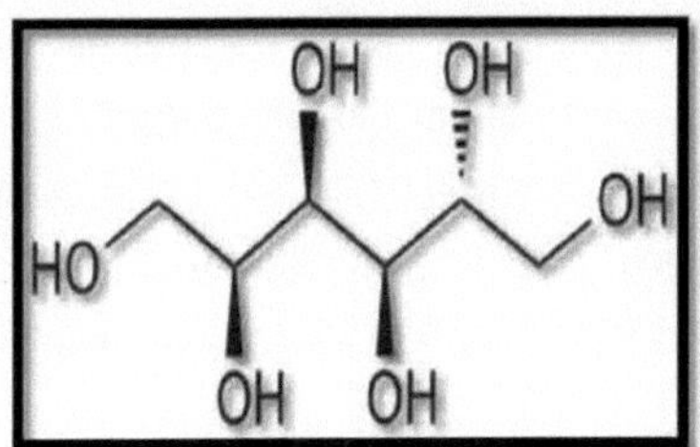

MANNITOL

A polyol sugar alcohol mannitol was originally isolated from the secretions of theFlowering Ash, called Manna after their resemblance to the biblical food. It issimilar to xylitol and sorbitol chemically.Mannitol is used as a sweetener for people with diabetes, and is commonly used as asweetener in breath freshening candies as it has a cooling effect. It is about 50 percent assweet as sucrose. It does not promote tooth decay and has a low caloric content. Mannitoldoes not pick up moisture and for this reason it is often used as a dusting powder for chewinggum. Due to its high melting point, it is also used in chocolate flavored coating agents for ice-cream and sweets.

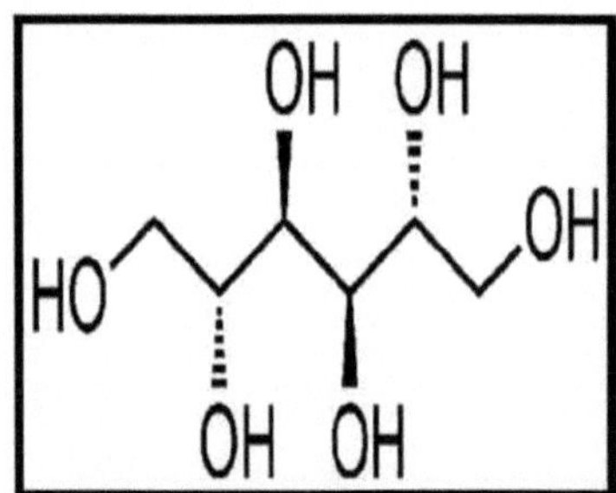

ERYTHRITOL

Erythritol like sorbitol and xylitol is a natural sugar alcohol derived from fruits and vegetables and approved for used as a food. Erythritol is produced commercially by fermentation of glucose. Erythritol is about 60-80% as sweet as sucrose and has a calorie value of 0.2 calories per gram. It is used primarily in chewing gum, baked goods and beverage and occurs naturally in pears, soy sauce, watermelon and grapes. In fact, Erythritol has even been found to exist naturally inhuman tissues and body fluids. Erythritol does not promotetooth decay and does not cause gastric side effects like other sugar alcohols.

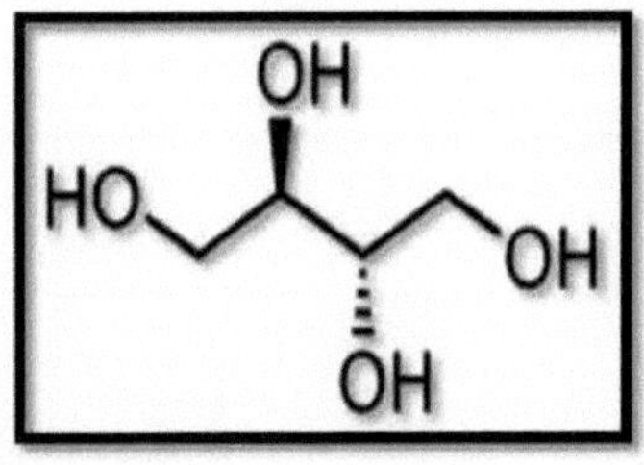

NATURAL SWEETENERS

Some natural sweeteners are also used as sugar substitutes because of their healthier benefits as compared to the processed table sugar or other sugar substitutes. Stevia is an important sweetener considered to be natural. Some other examples of such natural sweeteners include Mogrosides, Thaumatin, Brazzein, Date sugar, Grape juice concentrate, Honey, Maple sugar, Maple syrup, Molasses, and Agave nectar are safe natural sugar.

STEVIA

Stevia is used as sweetener but as it has a plant origin therefore Stevia is considered as natural sweetner. FDA also recommended Stevia as generally recognized as safe (GRAS) for use as general purpose sweeteners in 2008 and no questions regarding its safety mainly due to its plant origin. Stevia is basically highly purified steviol glycosides extracted from the Stevia plant. Stevioside a phytochemical found in Stevia plant is 300 times sweeter than sugar. Stevia sweeteners are natural, contain zero calories and are 200-300 times sweeter than sugar. Stevia sweeteners are approved for food and beverage use in several countries and can be found in the U.S. in many food and beverage products, including some juice and tea beverages, as well as some tabletop sweeteners. Its use as a natural sweetener and sugar substitute is frequent for diabetics, high blood pressure, cavity prevention, as a weight loss aid. It also shows antibacterial, antifertility, anti-inflammatory, antiseptic properties. It has similarly digestive tonic properties and also shows good results in cleaning up skin difficulties like acne, seborrhea, dermatitis, eczema etc.

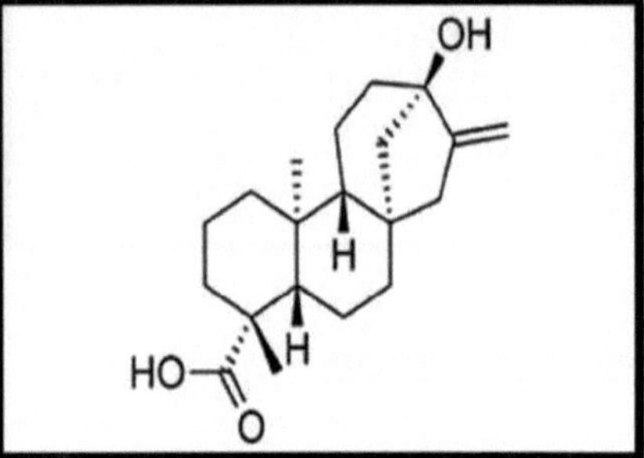

MOGROSIDES

Mogrosides are natural sugar substitutes extracted from the plant monk fruit (*Siraitiagrasvenorii*). Monk fruit plant has been cultivated and consumed in China for hundreds of years. Mogrosides are basically formed of varying numbers of glucose units from 2 to 6 attached to the mogroside unit and exist in the five chemical structures of mogrosides(II,III,IV,V and VI).Mogrosides are Generaly Recognized as Safe (GRAS) by FDA. It is about 100-250 times sweeter than sucrose and the sweeter level vary depending on the percentage of mogroside VI and the application/formulation. It has a pure and clean sweet taste, soluble in water, heat stable up to 125^{0}C, non-nutritive and with zero calorie.It has no limit Acceptable Daily Intake (ADI) Mogrosidesare available in the market in the form of solid or liquid under trade name ***Purefruit***. It has wide applications as sweetener and flavor for different food products, beverages and powdered drinks, chewing gums, baked goods, dietary supplements, nutritional bars and chocolates.

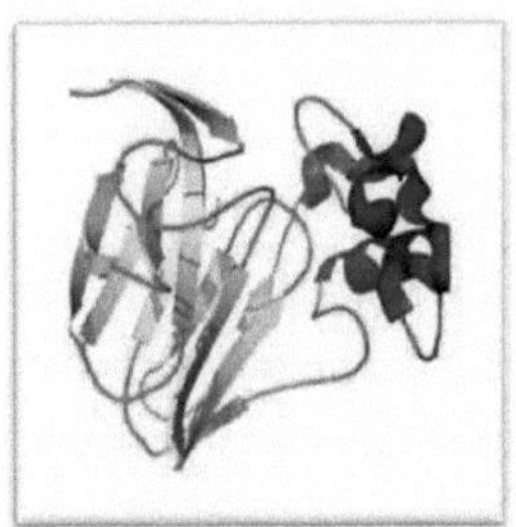

Structure of Mogroside V

THAUMATIN

Thaumatin is a low calorie protein sweetener and flavor enhancer extracted from West African fruit (katemfe fruit) *Thaumatococcusdanielli.* The plant has basically two types of amino acids Thaumatin IIprecursor for Thaumatin I which is the sweetener. It is about 2000-3000 times sweeter than sucrose and metabolized by the body as any other protein. It is a Generally Recognized as Safe (GRAS) by FDA in USA and also gained approval for over 30 countries around the world. Thaumatin is natural sweetener in a dried form available in the market under the trade name Talinand has no limit Acceptable Daily Intake (ADI). It is stable in freezing temperature, heat, and pH and soluble in water. Also combined with other low-calorie sweetener to obtain good sweetening combinations. It is widely used in food and beverages, sweetener blends, pharmaceutical and vitamin tablets, oral care products, animal feed and pet foods. Some limitations include delay perception of sweetness levels and leaving an aftertaste at high usage levels.

BRAZZEIN

Brazzein is basically a sweet tasting peptide extracted from west African fruit *Pentadiplandrabrazzeana.* It structure comprises of 54 amino acid arranged in one alpha helix and three strands beta-sheets. Although its large scale extraction yield from the fruit is notfeasible, but it has been genetically engineered in corn. The gluten protein from the modified corn contains 4% brazzein. It is non-caloric sweetener and about 1200 times sweeter than sucrose. Its taste is similar to sucrose with lingering sweet after taste. It is stable at the pH range of 2.5-8.0, and heat stable at 98^{0}C making itsuse in many commercial applications. It is commercially available in small packets under the brand name Cweet.

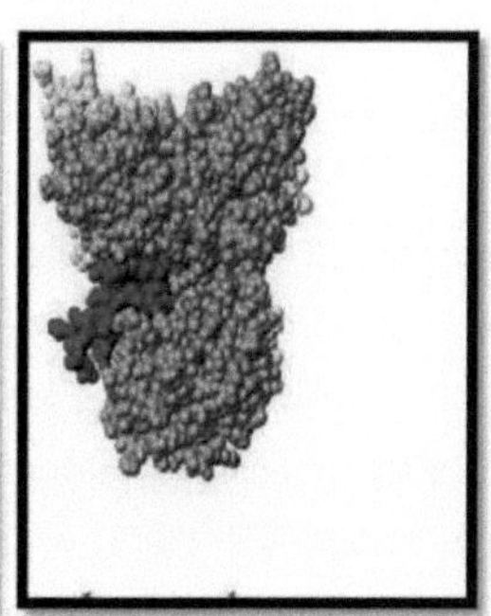

METABOLISM OF ARTIFICIAL SWEETNERS

It is estimated via various studies that sweetened food are source of weight gain because they induce sensation of appetite, hence increase in food intake takes place. Both natural as well as artificial sweeteners stimulates production of taste by acting on lingual taste bud as well as sugar digesting molecule by acting on GLP-I releasing cell of intestinal mucosa (Jang *et al.*, 2007). Variation in GLP-I secretion and gastric emptying was observed in those young healthy adult who consume diet soda before intake of glucose (Brown & Rother, 2009).

Their analysis suggest that NAS have increased GLP-I secretion which reduce absorption time of sugar and hence insulin secretion is effected in antagonistic manner hence appetite and glucose variation can takes place (Brown *et al.*, 2010). A study suggest that use of Soda/carbonated drink that are main source of artificial sweetener in children which induce overeating and results in weight gain in those children (Hill, 1965). Urine analysis shows that NCAS are excreted through body without being metabolized hence are non-caloric (Nabors, 2012).

ACCEPTABLE DAILY INTAKE

Acceptable daily intake ADI is basically an intake which individual is exposed to daily over its lifetimes without appreciable health risks (World Health Organization, 2004, p. 10). For the use of artificial sweeteners there are also standards and regulations set by the United States and other countries and same for their ADIs. ADIs for artificial sweeteners are calculated according to current safety research and can change with the advent of additional research warranting a reduction or complete removal of the product from the market. An example of such case is Cyclamate which was banned from the U.S. market in 1970 after several experimental studies on rats demonstrated its carcinogenic potential.

The ADI for saccharin is currently 5 mg/kg of body weight per day. The ADI for acesulfame-K for both the FDA and the Joint Expert Committee for Food Additives (JECFA) is 15 mg/kg of body weight per day. The European Union (EU) reevaluated this sweetener and supports its safety, but recommended an ADI of 9 mg/kg/d. The ADI for sucralose is 5 mg/kg/d. In 2002, the FDA set the ADI for neotame at 18 mg/kg/d. The JECFA confirmed the safety of neotame in 2003 and granted an ADI of 2 mg/kg/d (American Dietetic Association, 2004). Aspartame has an ADI of 50 mg/kg/d in the United States and 40 mg/kg/d in the EU (Soffritti et al., 2007).

World Production:

The world production of sugar and sweeteners was in 2002/2003 165,7 million tons sugar equivalents. 82 % is covered by sugar (142,6 Million tons), 11,5 % by starch sweeteners and sugar alcohols (17,6 Million tons) and 7 % by intense sweeteners 11,6 Million tons). Figure shows the actual production.

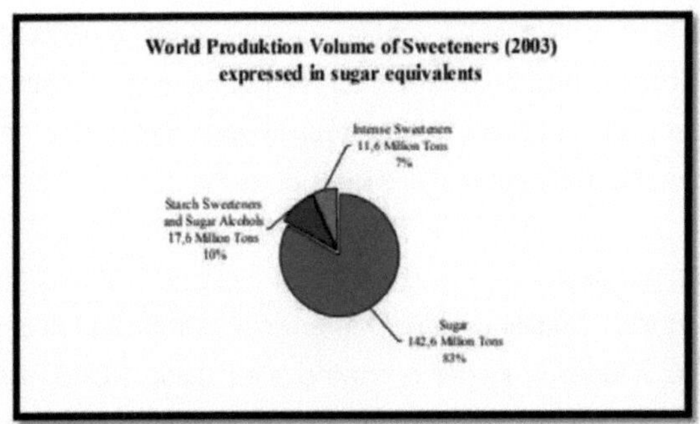

The total sugar and sweetener market represents a production volume of 49,6 billion US-$. Sugar is accounting for 75,2 % of production, starch sugars and sugar alcohols for 22,0 % and intense sweeteners only for 2,8 %.

BENEFICIAL EFFECTS OF ARTIFICIAL SWEETENERS

Artificial sweeteners are beneficial in that they provide sweetness, increasing the palatability of foods without the added sugar and resulting calories, an important adjunct to weight loss and diet programs. Most artificial sweeteners are not metabolized by the body and are therefore considered safe. However, scientists disagree about safety because the metabolites of the "non-metabolized" compounds have been shown to produce deleterious effects in mice, rats, and dogs.

HEALTH BENEFITS OF ARTIFICIAL SWEETENERS

PREVENTION OF DENTAL DECAY

Dental decay or dental caries is problem that takes place by the excess use of carbohydrates as carbohydrates act as substrate for acid releasing bacteria, this acid than results in dental decay. This dental decay process is done when natural sugars are used while non-nutritive sugars don't act as substrate hence acid production decrease (Charu *et al.*, 2012).

Therefore unlike sugar, artificial sweeteners don't contribute to tooth decay. The use of some artificial sweeteners will not harm your teeth and even some artificial sweeteners are designed to help rebuild your teeth. This can be incredibly beneficial for the people suffering with problems of teeth.

MAINTENANCE OF WEIGHT

Short term studies indicate that artificial sweeteners help in maintenance of body weight as it doesn't increase craving after using NCAS. One of the most appealing aspects of artificial sweeteners is that they are non-nutritive they have virtually no calories. In contrast, each gram of regular table sugar contains 4 calories. A teaspoon of sugar is about 4 grams. For perspective, consider that one 12-ounce can of a sweetened cola contain 8 teaspoons of added sugar, or about 130 calories. If you're trying to lose weight or prevent weight gain, products sweetened with artificial sweeteners rather than with higher calorie table sugar may be an attractive option. On the other hand, some research has suggested that consuming artificial sweeteners may be associated with increased weight, but the cause is not yet known.

Similarly all sugar alcohols such as xylitol, sorbitol, and mannitol have a lower caloric value than sugars. In this way, they can help people to achieve their weight goals. They are incompletely absorbed by the body and what is absorbed is metabolised by insulin-independent mechanisms orexcreted via the urine. A significant amount of what is not absorbed is metabolised to shortchain fatty acids and gases in the large intestine. Xylitol has 2.4 calories per gram; sorbitol 2.6 calories per gram; and mannitol 1.6 calories per gram. This is compared to the traditional 4 calories per gram that sugar has. All sugar alcohols have a low glycemic index, and they can be used to completely or partially replace traditional sugars such as sucrose and glucose. This helps to reduce the overall glycemic load of the diet, thus assisting with weight loss.

NCAS AND DIABETIC PATIENTS

Artificial sweeteners may be a good alternative to sugar for person with diabetes. Unlike sugar, artificial sweeteners generally don't raise blood sugar levels because they are not carbohydrates. (Butchko *et al.*, 2002). Sugars are naturally occurring carbohydrates. These include brown sugar, cane sugar, confectioners' sugar, fructose, honey, and molasses. They contain calories and raise your blood glucose levels. Reduced calorie sweeteners are sugar alcohols. The FDA has approved the low calorie sweeteners for people with diabetes, and the American Diabetes Association also recommends them. The product does not contain sugar at all, though it may contain sugar alcohols or artificial sweeteners. During processing, no extra sugar was added. However, the original source might have contained sugar, such as fructose in fruit juice. Additional sweeteners such as sugar alcohols or artificial sweeteners also might have been added. But because of concerns about how sugar substitutes are labeled and categorized, consultation with a doctor or dietitian about using any sugar substitutes is necessary if you have diabetes.

HEALTH RISKS OF ARTIFICIAL SWEETENERS

Many experimental studies were carried out to assess the deleterious effects of artificial sweeteners on health. Although such experimental designs to check the effects of NCAS on health is not an easy task as dietary composition changes by using these substitutes instead of sugar such as the nature of carbohydrate may change, fat, protein and carbohydrate proportions may change. Most of the researches regarding NCAS were carried out in animal models and toxicological studies of different artificial sweeteners have been done but still there are many discrepancies which limit the clear evidences regarding safety of these sweeteners. For example trials for observational studies in humans were mostly focused regarding the consumption of diet soft drinks as a replacement to regular soft drinks. Assessment for the association of the use of NCAS from sources other than diet soft drinks has been limited methodologically. Here is a brief review of the studies regarding health risks of some commonly used artificial sweeteners.

The Potential Toxicity of Artificial Sweeteners

Since their discovery, the safety of artificial sweeteners has been controversial. Artificial sweeteners provide the sweetness of sugar without the calories. As public health attention has turned to reversing the obesity epidemic in the United States, more individuals of all ages are choosing to use these products. These choices may be beneficial for those who cannot tolerate sugar in their diets (e.g., diabetics). However, scientists disagree about the relationships between sweeteners and lymphomas, leukemias, cancers of the bladder and brain, chronic fatigue syndrome, Parkinson's disease, Alzheimer's disease, multiple sclerosis, autism, and systemic lupus. Recently these substances have received increased attention due to their effects on glucose regulation. Occupational health nurses need accurate and timely information to counsel individuals regarding the use of these substances. This article provides an overview of types of artificial sweeteners, sweetener history, chemical structure, biological fate, physiological effects, published animal and human studies, and current standards and regulations.

TOXICOLOGICAL EFFECTS

Table 1: Potential toxicity of artificial Sweeteners (Whitehouse *et al.*, 2008).

			Manifestations of Toxicity in Humans	
Common Name	***Known Metabolites***	***ADI (mg/kg/d)***	***Acute***	***Chronic***
Acesulfame-K		15	Headache	Clastogenic, genotoxic at high doses, thyroid tumors in rats
Aspartame	Methanol, aspartic acid, phenylalanine	50	Headache, dry mouth, dizziness, mood change, nausea, vomiting, reduced seizure threshold, thrombocytopenia	Lymphomas, leukemias in rats
Cyclamate	Cyclohexylamine	1		Bladder cancer in mice, testicular atrophy in mice
Neotame	De-esterified neotame, methanol	2	Headache, hepatotoxic at high doses	Lower birth rate, weight loss (due to decreased consumption at higher doses)
Saccharin	O-sulfamoylbenzoic acid	5	Nausea, vomiting, diarrhea	Cancer in offspring of breast-fed animals, low birth weight, bladder cancer, hepatotoxicity
Sucralose		5	Diarrhea	Thymus shrinkage and cecal enlargements in rats

ADI = acceptable daily intake.

HEALTH ISSUES OF SACCHARINE

The acute problems regarding to saccharin include diarrhea and vomiting while long term exposure leads to cancer in offspring through breast feeding, low birth rate, bladder cancers and liver toxicity (Whitehouse *et al.*, 2008). Longitudinal studies on its health effects suggest that it is source of bladder cancers in male if consumed daily. Exposure studies of saccharin provide both positive and negative results, including the potential to induce cancer in rats, dogs, and humans. Arnold in 1983 discussed in a review about two-generation saccharin bioassays. In this study animals were exposed to saccharin, at all stages of development (i.e., in utero, during lactation, and in feed as an adult). According to these studies it was clearly demonstrated that when rats were exposed to diets containing 5% or 7.5% saccharin from the time of conception to death, an increased frequency of urinary bladder cancers was found, predominantly in males. It was noted that saccharin is not metabolized by the body as it is nucleophilic and does not bind to DNA. However, it does suppress humoral antibody production in rats.

Then further studies of saccharin were performed and resulted in the prohibition of saccharin in US and Canada. It was also required that all food containing saccharin bear the warning labelindicating that"saccharin is a potential cancer causing agent." Subsequent studies in rats showed an increased incidence of urinary bladder cancer at high doses of saccharin, especially in male rats. However, mechanistic studies (studies that examine how a substance works in the body) have shown that these results apply only to rats. Human epidemiology studies (studies of patterns, causes, and control of diseases in groups of people) have shown no consistent evidence that saccharin is associated with bladder cancer incidence.

Because the bladder tumors seen in rats are due to a mechanism not relevant to humans and because there is no clear evidence that saccharin causes cancer in humans, saccharin was delisted in 2000 from the U.S. National Toxicology Program's *Report on Carcinogens*, where it had been listed since 1981 as a substance reasonably anticipated to be a human carcinogen (a substance known to cause cancer). Furthermore this delisting led to legislation, which was then signed into law canceling the warning label requirement for products containing saccharin. Weihrauch and Diehl reported about more than 50 studies of saccharin in laboratory rats. They indicated that none of the groups in the first-generation rats demonstrated significantly more neoplasms in the saccharin-fed animals over controls. Whether according to a study by Fukushima et al. showed an increase in bladder neoplasms but continued to refute their results, stating the rats used in this trial were frequently infected with a urinary parasite that could have made the rats susceptible to bladder cell proliferation (Weihrauch& Diehl).

A case study of the hepatotoxicity of saccharin was published in 1994 (Negro, Mondardini, & Palmas). A patient presented with elevated serum concentrations of liver enzymes after the oral administration of three different drugs, of which saccharin was the only common constituent. Re-exposure to pure saccharin supported its role in pathogenesis of the liver damage.

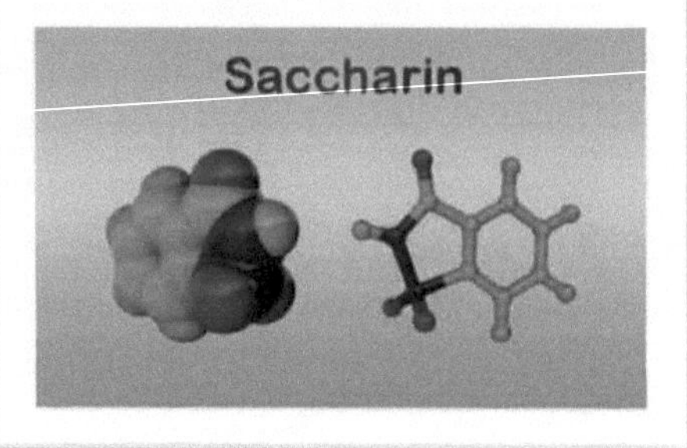

Evidence suggests that it's safe

Health authorities agree that saccharin is safe for human consumption. These include the World Health Organization (WHO), the European Food Safety Authority (EFSA), and the Food and Drug Administration (FDA). However, this wasn't always the case, as in the 1970s, several studies in rats linked saccharin to the development of bladder cancer (1Trusted Source).

It was then classified as possibly cancerous to humans. Yet, further research discovered that the cancer development in rats was not relevant to humans. Observational studies in humans showed no clear link between saccharin consumption and cancer risk (2Trusted Source, 3Trusted Source, 4 Trusted Source). Due to the lack of solid evidence linking saccharin to cancer development, its classification was changed to "not classifiable as cancerous to humans (5Trusted Source)." However, many experts feel observational studies are not sufficient to rule out that there's no risk and still recommend that people avoid saccharin.

We first investigated whether a four-week neotame consumption would affect the gut microbiome of CD-1 mouse. No significant difference of alpha-diversity was observed between two groups before neotame consumption. After the four-week experiment, alpha-diversity of gut microbiome in neotame-consuming group was much lower than the control group, as shown in Figure 1A. PCoA analysis showed a separation of gut bacteria between control and neotame-consuming animals after the four-week treatment, compared to their clustering distribution before neotame consumption (Figure 1B). The results suggest that neotame consumption significantly altered both the alpha- and beta-diversities of gut bacteria of mice. Dysbiosis analysis found that neotame-treated mice had a significantly higher microbial dysbiosis index (MD-index) than controls (Figure 2A). Specifically, phylum Bacteroidetes was largely enriched, while Firmicutes was significantly decreased in neotame-treated animals (Figure 2B). Before neotame treatment, no such significant taxonomy difference was observed on the phylum level. On genus levels, we found that *Bacteroides* and an undefined genus in family S24-7 mainly contributed to the increase of phylum Bacteroidetes, as shown in Figure 2C.

Over 12 genera have been significantly altered in Firmicutes (Table S1). Notably, multiple components of family Lachnospiraceae and family Ruminococcaceae in neotame-treated animals were significantly lower controls,suchas *Blautia*, *Dorea*, than *Oscillospira* and *Ruminococcus* (Figure 2D,E). More altered genus can be found in Table S1 (Supplementary Materials).

Taken together, the results suggested that the four-week neotame consumption perturbed the diversities as well as the community compositions of gut microbiome in male CD-1 mice.

SUMMARY

Observational studies in humans have found no evidence that saccharin causes cancer or any harm to human health.

Food sources of saccharin

Saccharin is found in a wide variety of diet foods and drinks. It's also used as a table sweetener. It's sold under the brand names *Sweet 'N Low, Sweet Twin, and Necta Sweet.* Saccharin is available in either granule or liquid form, with one serving providing sweetness comparable to two teaspoons of sugar. Another common source of saccharin is artificially sweetened drinks, but the FDA restricts this amount to no more than 12 mg per fluid ounce. Due to the ban on saccharin in the 1970s, many diet drink manufacturers switched to aspartame as a sweetener and continue to use it today.

Saccharin is often used in baked goods, jams, jelly, chewing gum, canned fruit, candy, dessert toppings, and salad dressings. It can also be found in cosmetic products, including toothpaste and mouthwash. Additionally, it's a common ingredient in medicines, vitamins, and pharmaceuticals. In the European Union, saccharin that has been added to food or drinks can be identified as E954 on the nutrition label.

HEALTH ISSUES OF ASPARTAME

Aspartame has been considered the most controversial artificial sweetener because of its potential toxicity. Aspartame was approved in 1981 by the FDA after numerous tests showed that it did not cause cancer or other adverse effects in laboratory animals. But the questions regarding the safety of aspartame were raised by a 1996 report as the number of brain tumor cases increased and this was associated with the use of aspartame.

However, an analysis regarding this issue showed that this problem of incidence of brain cancers began 8 years prior to the approval of aspartame. Moreover, increases in overall brain cancer incidence occurred primarily in people age 70 and older, a group that was not exposed to the highest doses of aspartame since its introduction. These data do not establish a clear link between the consumption of aspartame and the development of brain tumors. In 2005, a laboratory study found more lymphomas and leukemias in rats fed very high doses of aspartame (equivalent to drinking 8 to 2,083 cans of diet soda daily).

Soffritti and coworkers in 2007 established positive link between aspartame intake and different carcinomas and malignant tumors in both rat and human i.e. leukemia in both male and female, mammary glands cancers in females and various lymphomas including NHL and B-lymphocytic lymphoma etc. carcinogenic activity of aspartame increase if its intake is greater during gestation period. It is also observed that excess use of aspartame can leads to migraine, decrease in platelets, enlargement of spleen and liver (Whitehouse *et al.,* 2008).

Blumenthal (1997) reported three case studies wherein women ages 40, 32, and 26 all experienced migraines while chewing a popular gum with aspartame additive. In all cases, the migraines were relieved after cessation of product use. The headaches were reproducible by reintroducing the gum. Additionally, a case report in 2007 revealed four individuals with thrombocytopenia attributed to products containing aspartame (Roberts, 2007) Hypothalamic neuronal necrosis due to dicarboxylic amino acids (i.e., the aspartic acid found in aspartame) was observed in a newborn rodent model,. However, this necrosis was not observed in a non-human primate model, even at 10-fold higher doses (Stegink, Shepherd, Brummel, & Murray, 1974; Stegink, 1976). The other amino acid found in aspartame, phenylalanine, is metabolized differently in rodents, so only data from primate animal models can be considered. Phenylalanine doses of 3,000 mg/kg/d in the first few years of life produced irreversible brain damage in monkeys (Waisman& Harlow, 1965). Unfortunately, no dose-response data are available for use in estimating human effects.

In humans, aspartame doses of 2 to 100 mg/kg resulted in dose-related increases in phenylalanine without an observed effect on behavior or cognitive performance (Filer &Stegink, 1988; Lieberman, Caballero, Emde, & Bernstein, 1988; Stokes, Belger, Banich, & Taylor, 1991). In sub-chronic dosing studies, aspartame doses of 30 to 77 mg/kg/d over 13 weeks in 126 children and adolescents and a similar study in young adults administered 36 mg/kg/d showed no significant impact on renal or hepatic function, hematologic status, ophthalmic examinations, or plasma lipid profile (Frey, 1976; Knopp, Brandt, &Arky, 1976).However, there were some inconsistencies in the findings. For example, the number of cancer cases did not rise with increasing amounts of aspartame as would be expected. In an another study Increasing consumption of aspartame-containing beverages was not associated with the development of lymphoma, leukemia, or brain cancer.

By far, aspartame has been the most controversial artificial sweetener because of its potential toxicity. Numerous websites are devoted to removing aspartame from all sources immediately. Although some of these websites list relevant literature and demonstrate cause and effect, others attribute an all-inclusive disease list to the ingestion or absorption of aspartame. New research from Soffritti *et al.* (2007) provides evidence of the carcinogenic potential of this compound. Their research, using Sprague Dawley fetal rats, has demonstrated a significant increase of malignant tumors in males, an increase in the incidence of lymphomas and leukemias in males and females, and an increase in the incidence of mammary cancer in females. These results reinforce and confirm previous research that also demonstrated the carcinogenicity potential of aspartame and the increased carcinogenetic potential if exposure occurs during gestation. It is notable that the dosage tested approximated the ADI for humans. In other published reports, Blumenthal (1997) reported three case studies wherein women ages 40, 32, and 26 all experienced migraines while chewing a popular gum with aspartame additive. In all cases, the migraines were relieved after cessation of product use. The headaches were reproducible by reintroducing the gum. Additionally, a case report in 2007 revealed four individuals with thrombocytopenia attributed to products containing aspartame (Roberts, 2007).

This conclusion was based on recurrence of blood dyscrasia on two or more occasions after rechallenge, and the absence of any other definable factors. One of the reports was of a 10-year-old girl who developed a decline in platelet count to 1,000 cu/mm, coupled with enlargement of the liver and spleen, and a marked increase in histiocytes in the bone marrow. A dramatic clinical and hematological normalization followed when additives were eliminated from her diet. Similar recurrences were documented twice after ingesting aspartame.

Remissions were maintained when the client abstained from aspartame products (Roberts). In a newborn rodent model, hypothalamic neuronal necrosis due to dicarboxylic amino acids (i.e., the aspartic acid found in aspartame) was observed. However, this necrosis was not observed in a non-human primate model, even at 10-fold higher doses (Stegink, Shepherd, Brummel, & Murray, 1974; Stegink, 1976). The other amino acid found in aspartame, phenylalanine, is metabolized differently in rodents, so only data from primate animal models can be considered. Phenylalanine doses of 3,000 mg/kg/d in the first few years of life produced irreversible brain damage in monkeys (Waisman & Harlow, 1965). Unfortunately, no dose-response data are available for use in estimating human effects.

HEALTH STUDY ON ACE-K

Longitudinal study of Ace-K on animal and mice revealed that there is no role of Ace-K on cancer development and mutation in DNA if the daily intake is 0-9 mg/Kg. If the intake increases from that value it will result in increase in body weight and may be act as DNA damaging agent. Ace-K can be used to treat neurodegenerative disorders because its excess use results in decrease in neuro-protective activity of cell by inhibiting ATP production(Chadwick W, 2010). Evidence suggests that consumption of Ace-K induce increase in plasma insulin level. Acute symptoms related to Ace-K are headache, dizziness, nausea and vomiting, while long term exposure leads to leukemia as well as lymphomas (Whitehouse *et al.*, 2008).

A cytogenic study of this sweetener indicated that when the dosage administered to mice was within the ADI of 15 mg/kg of body weight, the number of chromosomal aberrations was not significant compared to control mice. However, at higher doses (60, 450, 1,100, and 2,250 mg/kg), acesulfame-K was clastogenic and genotoxic. Therefore, the results demonstrated that genetic damage caused due to interaction of acesulfame-K with DNA is dose dependent. Additional studies on genotoxicity are recommended because of potential DNA interaction at high doses. However, doses capable of producing damage are well above the ADI.

One of the most serious claims about acesulfame potassium is that it could increase the risk of cancer. In 1996, the Center for Science in the Public Interest (CSPI) openly queried the quality of the science used to approve this sweetener for widespread use (15). However, the FDA and the NCI say that acesulfame potassium is safe and that there is enough evidence to say that it does not cause cancer (2, 4). Scientists have tested whether acesulfame potassium could cause cancer using both test tubes and animals. In test tubes, they look for signs that a substance could be "genotoxic," which means that it could damage DNA and cause mutations that may lead to cancer. Many studies have failed to detect any signs of genotoxicity.

In 2005, the National Toxicology Program conducted one of the largest animal studies. They gave mice up to 3% of their total diet as acesulfame potassium for over 40 weeks- the equivalent to a person drinking more than 1,000 cans of soft drinks each day. They found no evidence of an increased risk of cancer in the mice (16). In summary, studies in test tubes and lab animals suggest that acesulfame potassium does not cause cancer. Although some disagree, major regulatory authorities have reached the same conclusion.

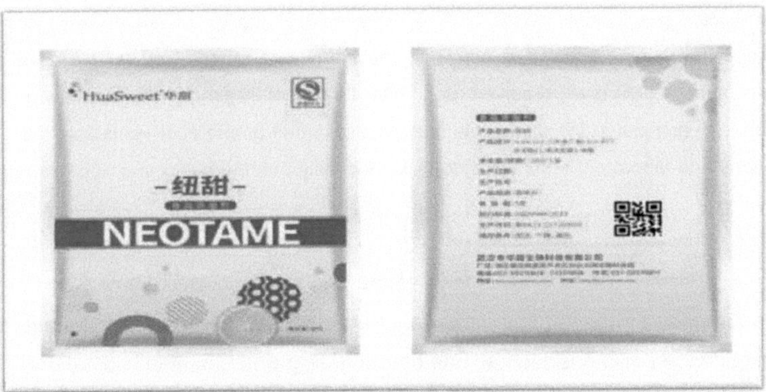

HEALTH ISSUES OF NEOTAME:

Short term study on health effects of Neotame suggests that there is no effect of Neotame dose during pregnancy and fetus development, and genetic variation. The only issue with this sweetener is that there is no labeling on this for those who are suffering from phenylketonuria (Nabors, 2012). Studies of neotame reveal changes in body weight, body weight gain, and food consumption. It is noted that these effects are not due to the toxic profile of neotame, but rather to the unpalatability of feeds containing this sweetener. Therefore, rats will decrease their daily food intake, resulting in long-term loss in body weight and less weight gain.

In definitive safety studies, no adverse findings related to neotame treatment were found in clinical observations, physical examinations, water consumption, or clinical pathology evaluations; nor were there morbidity, mortality, organ toxicity, or macroscopic or microscopic postmortem findings (Mayhew, Comer, & Stargel, 2003). The study was conducted to evaluate tolerance of a single dose of neotame ingested in solution by healthy men. The safety of neotame was evaluated in a stepwise fashion, starting with the lowest dose, followed by the intermediate dose and then the high dose only after safety at lower doses had been confirmed. Nineteen healthy men (mean age ±SD, 28 ±6 years) were given single doses of 0.10, 0.25, or 0.50 mg/kg bw (seven, six, six men per dose, respectively) of neotame in solution. The study was randomized, single dose and not double-blinded. Each man received only one treatment regimen after an overnight fast. Eighteen men completed the study; one man was excluded due to poor venous access. Clinical evaluations and laboratory tests were done immediately before dosing and approximately 48 h after dosing.

There were no treatment-related changes in pulse rate or blood pressure, and no changes in haematology, clinical chemistry or urine analysis parameters. Two men experienced mild headaches, one before dosing and one after a dose of 0.1 mg/kg bw per day. At 0.5 mg/kg bw per day, another two men had mild headaches, one before dosing and one after, and one had lower back pain. These signs resolved without further treatment and were not attributed to dosing with neotame (Kisicki et al., 1997).

HEALTH ISSUES OF SUCRALOSE

Toxicology studies of sucralose show little effect, the most significant finding being shrunken thymus glands with diets of 5% sucralose. Review and evaluation of the data, including a special immunotoxicity study, clearly demonstrated that thymic changes were not a manifestation of intrinsic toxicity, but rather an exacerbation of the normal process of involution resulting from a nutritional deficit to reduced food intake (Grice & Goldsmith, 2000). Cases studies have been reported on sucralose consumption and the increased incidence of deleterious effects.

Bigal and Krymchantowski (2006) discuss migraines triggered by sucralose. Multiple blinded posttests were provided for this client once she was migraine free to determine if sucralose was the trigger for her migraines. In all cases, on consumption of sucralose, her migraines returned. The client refused further participation in the study due to the migraines associated with the sucralose-containing test solution. McLean Baird, Shephard, Merritt, and Hildick-Smith (2000) published a study of sucralose tolerance in humans. Sucralose was administered to eight individuals for up to 12-week intervals. They experienced no adverse health effects at doses up to 10 mg/kg/d and repeated doses increasing to 5 mg/kg/d for 13 weeks. Although an important study, it is limited by the number of participants and the length of post intervention study.

Sucrose administration increased alanine aminotransferase (ALT) and aspartate aminotransferase (AST) levels in male mice groups III and IV in a significant way; while no significant change was observed upon their administration in female mice groups (Table 3, Fig. 2). Short and long term administration of sucralose, in both male and female mice groups, significantly elevated ALT and AST levels when compared to their corresponding control groups or sucrose administrated groups III and IV. Stevia administration for 8 weeks significantly increased ALT level, in both male (48.44 ± 9.5) and female (48.44 ± 9.1) mice group VII; this increase continued to reach the highest level among all experimental groups (57.14 ± 3.6 and 56.14 ± 2.6 for male and female mice group VIII, respectively). AST level, in stevia administrated female group VII (42.21 ± 1.6) and VIII (49.08 ± 0.7), was significantly higher than that of control groups I and II (25.62 ± 2.5 and 26.55 ± 19.4) and sucrose administrated groups III and IV (29.30 ± 3.2 and 31.31 ± 0.4, respectively); but significantly lower than that of sucralose administrated groups (52.01 ± 5.2 and 51.34 ± 0.8 for group V and VI, respectively).

While there aren't a significant number of studies pertaining to the impact of regular intake of artificial sweeteners on reproductive health and fertility outcomes, the existing evidence points to a cause for concern and further research. [2]

Preclinical mechanistic studies have observed a range of physiological effects with oral exposure to artificial sweeteners, particularly saccharine, sorbitol and aspartame. In female animal models these artificial sweeteners were shown to significantly increase the percentage of abnormal estrous cycles, serum progesterone levels and apoptosis in granulosa, oocyte and corpus luteum cells, and adversely influence the quality of human oocytes and embryo development. [2,9] In pregnant animal models, artificial sweeteners resulted in significant decreases in circulating progesterone levels, placenta diameter and embryo glutathione concentrations, induced glucose-intolerance and a pro-oxidative environment in the endometrium. The adverse influence of artificial sweeteners on glucose is independent of their low-caloric status. [2,10] Consequently, offspring exposed to artificial sweeteners in utero had reduced gestational lengths and foetal weights and increased chromosomal aberrations, foetal congenital malformations and embryo toxicity and decreased embryo development scores.

The latter mechanism has been partly attributed to the impact of the aspartame metabolite formic acid and is one of the mechanism suggested to be a potential factor involved in the increased risk of preterm birth observed in human clinical trials. [2,10-13] Male animal models have also observed negative effects associated with exposure to artificial sweeteners on sperm parameters and an increased rate of sperm DNA fragmentation and apoptosis. [14] The importance of local taste receptors in sperm maturation in the concentration of fertilisation-competent sperm is one recent area of investigation which may be a mechanism influencing the impact of artificial sweeteners on sperm health. [15] Some of these physiological mechanisms associated with artificial sweeteners ingestion in animals have been also observed in human clinical trials. [2,16]

CONTROVERSIES ABOUT HEALTH EFFECTS OF ARTIFICIAL SWEETENERS

The promising role of artificial sweeteners whether beneficial, harmful or neutral is still a debatable point and a lot of controversies exist about their health effects. The truth behind these controversies is not clarified yet. People with diabetes and obesity are using these artificial sweeteners frequently and studies showed the positive results for controlling these epidemics. On the other hand, some studies have shown that products that contain artificial sweeteners can actually helpto promote weight gain. The phenomenon behind this study is that when these artificial sweeteners are eaten, body prepares itself for a calorie load but in actual this not happens that's why people either keep on eating to obtain the required calories or reduce calorie burning activity. In particular, very little data exist regarding the role of artificial sweeteners in glucose metabolism in children. Our growing understanding of the active metabolic role played by such chemicals in animal models should spur further research into the effects of these common food additives in humans.

Many of the studies on product safety have been conducted by companies that produce these products and are not generally available to consumers. A Medline search of publications from 2000 to 2008 using the key words "artificial sweeteners," "sweetening agents," "toxicity," "toxicology," "safety," and "consumer product safety" resulted in only one study available to readers of primary product safety data. Groups that believe the safety of these substances has not been demonstrated point to the length of studies, sample sizes, and lack of controls.It is currently recommended that foods and drinks that contain artificial sweeteners be used inmoderation. Do not use them as an excuse to indulge in other high calorie foods or to skipphysical activity that is important to weight control and health.

SUSCEPTIBLE POPULATIONS

People who are at high risk from the potent harmful effects of these artificial sweeteners include diabetics, children, pregnant women, women of childbearing age, breastfeeding mothers, individuals with low seizure thresholds, and individuals at risk for migraines.

More studies are required for these susceptible populations. A focus on children is important because they have a higher intake of foods and beverages per kilogram of body weight. Also, more research on the effect of artificial sweeteners on diabetic clients is needed because this population is likely toingest larger quantities of sugar substitutes. Because artificial sweeteners are in more than 6,000 products, including foods, medications, and cosmetics, it is impossible to completely eradicate them from daily encounters. Replication studies and long-term assays are required to decrease fear resulting from the limited research that currently exists.

IMPLICATIONS FOR PRACTICE:

As the rate of obesity and diabetes is increasing continuously and such people find artificial sweeteners a blessing to enjoy the taste of their foods. But it is necessary to keep in view all the facts related to such products if these are in daily use. There is a need of up-to-date and evidence based information to guide the people. Counseling of individuals in primary care settings, schools, and workplaces is of increasing importance.The use of non-nutritive sweeteners, when consumed in a diet must be guided by daily intake recommendations according to standard dietary guidelines.

And the individuals must be aware about the benefits of artificial sweeteners as well as potential deleterious side effects associated with their use. Much literature is available on artificial sweeteners; however, it is difficult for the general public to decipher the research, especially when researchers themselves disagree. Health care providers should be aware of current research surrounding the use of artificial sweeteners and inform clients of the potential risks associated with their use. The best possible way is to consume these substances in moderation or not at all, go for healthy and natural foods and involvement of physical activities in daily routine and use of such additives under the guidance of a health care provider.

DETAILED STUDY ON SACCHARIN

THE METABOLISM AND TOXICOLOGY OF SACCHARIN

➢ Origin of Saccharin:

Saccharin was serendipitously discovered in 1879 by Constantine Fahlberg, achemist of Johns Hopkins University as one of the first artificial sweeteners on earth. This happened when he was researching the oxidative mechanisms of toluenesulfonamide while working on coal tar derivatives in the laboratory of Ira Remsen. Accidentally, he spilled a chemical on his hand. Later on, while eating dinner, Fahlberg noticed a more sweetness in the bread he was eating, he licked his finger and noticed that the substance had a sweet taste [5, 6, 2,7]. Through careful examination, he traced the sweetness back to the chemical, later named saccharin, by tasting various residues on his hands and clothes and finally chemicals in the laboratory.

In 1879 and 1880, Fahlberg and Remsen published articles on benzoic sulfinide. Fahlberg described the methods of producing this substance that he named saccharin when he was in New York City working on his own in 1884 [5]. Since the time of saccharin discovery, a number of compounds have been discovered and used as food additives for their sweetening properties. Its use has been since 1900, but obtained FDA approval in 1970 [7]. By 1907, saccharin was used as a replacement for sugar in foods for diabetics. Since it is not metabolized in the body for energy, saccharin is classified as a non-caloric sweetener. By the 1960s it was used on a massive scale in the "diet" soft drink industry [2]. Consequent upon sugar shortage during the World War I, saccharin became widespread and commercialized. Since saccharin is a calorie-free sweetener, its popularity further increased during the 1960s and 1970s among dieters [2]. In the United States, saccharin is often found in restaurants in pink packets; the most popular brand is "Sweet 'N Low". Saccharin is used to sweeten products such as drinks, candies, medicines, and toothpaste, canned fruit, jams, salad dressing, chewing gum, table top sweeteners, baked goods, jams, and dessert toppings etc [2].

➢ Can everyone consume artificial sweeteners?

It can be difficult to determine if there will be any long-term health problems if you consume artificial sweeteners during pregnancy or during childhood. None of the manufacturers say that you can't consume them, but none of them have long-term studies that prove that it is safe to do so. Pregnancy is a time when every bite and sip that a woman takes really matters. There have been studies in rats that show when life span exposure to artificial sweeteners begins during fetal life, its carcinogenic effects are increased. The Danish National Birth Cohort is a study of 59,334 women conducted from 1996 to 2002. They found that daily intake of artificially sweetened soft drinks may increase preterm delivery.

There are numerous studies that show the negative health consequences of diets high in sugar for children. Many measures are being taken to help decrease the amount of sugar that children consume. This means that many people are turning to artificially sweetened foods and beverages for a child's sweet tooth. Unfortunately, there are no studies on the effects of these sweeteners and possible long-term consequences of children consuming them. When possible, it's best for pregnant women and children to avoid artificial sweeteners. If they are When possible, it s best for pregnant women and children to avoid artificial sweeteners. If they are used, they should be used in moderation. It's always important to weigh the overall health benefits when choosing foods and beverages. Consult your doctor if you have any questions or concerns about this.

Effects of artificial sweeteners on food intake in children

Several studies have looked at the effect of caloric versus artificial sweetener preloads on subsequent *ad libitum* food intake in children. In two experiments, boys aged 9–14 with a wide range of body mass indices (BMIs) exhibited complete caloric compensation in *ad libitum* lunch intake 30 minutes following a sucralose versus sugar-sweetened drink, meaning that they reduced their intake at lunch by the number of calories contained in the preload (44,45).

In another study, 2.5 to 5-year-old children showed complete caloric compensation 20 minutes after a low-calorie, aspartame-sweetened pudding preload versus high-calorie, maltodextrin-sweetened pudding (46). Interestingly, adults participating in the same experiment showed no caloric compensation, meaning they ate the same number of calories at lunch regardless of which preload they received.

The timing of artificial sweetener consumption with respect to meals may also affect food intake. Children who consumed sugar-sweetened beverages immediately prior to a test meal ate less than those who consumed artificially-sweetened beverages before the meal (47). However, no compensation in meal intake was observed when sugar- versus artificially-sweetened beverages were given 30, 60, or 90 minutes prior to the test meal (47,48). In addition, children reduced *ad libitum* lunch intake 30 minutes after an aspartame-sweetened preload (compared with water), but not after 0 or 60 minutes (47). These studies, of course, do not describe the effect of chronic consumption of artificial sweeteners on food intake. Birch et al. explored this idea by giving 3 to 5-year-old children an *ad libitum* snack 20 minafter either low-calorie (aspartame-sweetened) or high-calorie (maltodextrin-sweetened) pudding during multiple trials (49). With each trial, children consistently showed complete caloric compensation for the preload. However, when subsequently given an intermediate caloric density pudding of the same flavor they had received previously, children who had eaten low-calorie pudding beforehand ate significantly more snack than those previously given high-calorie pudding, by approximately 50 kcal.

Observational studies of artificial sweeteners and weight gain in children

The majority of pediatric epidemiologic studies have found a positive correlation between weight gain and artificially-sweetened beverage intake. Blum et al. examined beverage consumption and BMI Z-scores in 164 elementary school-aged children (43). This longitudinal study found that increased diet soda consumption was positively correlated with follow-up BMI Z-score after two years.

Comparable results were found by Berkey et al., who examined the relationship between BMI and diet soda consumption in over 10 000 children (aged 9 to 14 years) of Nurses' Health Study II participants over the course of one year (50). Artificially-sweetened beverage intake was significantly correlated with weight gain in boys, but not in girls, during the study period. A long-term prospective study of 1 203 children in England found that artificially-sweetened beverage consumption at ages 5 and 7 was correlated both with baseline BMI and fat mass at age 9 (51). Another longitudinal study of 2 371 girls (aged 9 and 10) participating in the National Heart, Lung and Blood Institute Growth and Health Study showed that diet soda consumption was significantly associated with higher daily caloric intake, but not with BMI (41). A much smaller study of 177 children aged 3 to 6 years showed no association between diet soda consumption and risk of obesity (42).

Several cross-sectional studies in children have added to the association between artificially-sweetened beverage use and adverse health effects. Forshee et al. analyzed data from the US Department of Agriculture's Continuing Survey of Food Intake by Individuals from 1994–96 and 1998. This nationally representative sample of US children between 6 and 19 years of age found that BMI was positively correlated with diet soda consumption (52). These results were consistent with Giammattei et al.'s findings in 385 sixth and seventh graders, which showed that both diet and sugar-sweetened soda intake were positively correlated with BMI z-score and percent body fat (53). However, a study of 2 to 5-year-old children using NHANES data did not show an association between artificially-sweetened beverage consumption and BMI in this age group (54).

To date, only one observational study has shown an inverse association between artificial sweetener use and weight gain. In this study of 548 ethnically diverse school children (mean age 11.7 years) in Massachusetts, Ludwig et al. found that increased diet soda consumption over a 19-month time period was associated with decreased incidence of obesity, whereas the odds ratio of becoming obese increased 1.6 fold for each sugar-sweetened drink consumed (55).

Interventional studies of artificial sweeteners and weight gain in children

Three small interventional studies that manipulated artificial sweetener intake have been conducted in children, and have failed to show metabolic effects. Shortly after the approval of aspartame by the FDA, its effects during active weight reduction and its role in glucoregulatory hormone changes were studied in 55 overweight children and young adults, aged 10 to 21, during a 13-week 1 000 kcal/day diet (56). There were no differences in weight loss for subjects receiving 2.7 g/day of encapsulated aspartame versus placebo.

A randomized, controlled pilot study of 103 adolescents, aged 13 to 18 years, examined the effect of replacing sugar-sweetened drinks with artificially-sweetened beverages or water during a 25-week period (57). Changes in BMI for intervention versus control (no replacement of sugar-sweetened drinks) were not significant for the entire group, although an exploratory *post-hoc* analysis showed that the intervention made the greatest difference in the heaviest subjects, whose BMIs declined by 0.63 ± 0.23 kg/m_2, compared with a 0.12 ± 0.26 kg/m_2 gain in the control group. However, the authors did not separately report consumption of water versus artificially-sweetened beverages during the intervention, and thus the effect of artificial sweeteners in this study could not be isolated.

In the third randomized, controlled trial, girls aged 11 to 15 years consumed a 1500 kcal/day diet for 12 weeks. In one group, sugar-sweetened soda was permitted as a snack, while in the other group, only diet sodas were permitted. There were no differences between groups for BMI change, and reported intake of either sugar-sweetened or artificially-sweetened soda did not affect BMI change (3).

REFERENCES

1. Arpe, H.-J. 1978. Acesulfame-k, a new noncaloric sweetener. In Health and Sugar substitutes. Proceedings of the ERGOB Conference, Geneva 1978, Ed. B Guggenheim. Basel: Karger. 178-183.
2. "Aspartame: Aspartame." Flavored Waters - Vitamin Enhanced Oxygenated and Structured Waters. Web. 20 Dec. 2009. <http://www.Flavored waters .Com/ sweeteners /aspartame.Asp>.
3. American dietetic association. 2014. Position of the american dietetic association: Use of nutritive and nonnutritive sweeteners. J. Am. Diet. Assoc. 104: 255-275.
4. Brown, R.J., B.A.D. Mary, R.I. Kristina. 2010. Artificial sweeteners: A systematic review of metabolic effects in youth. Int. J. Pediatr. Obes. 5(4): 305-312.
5. Brown Rj, W.M., K. Rother. 2009. Ingestion of diet soda before a glucose load augments glucagonlike peptide-1. Diab. Care. 32: 184-186.
6. Butchko, H.H., W.W. Stargel, C.P. Comer, D.A. Mayhew, C. Benninger, G.L. Blackburn, L.M. De Sonneville, R.S. Geha, Z. Hertelendy, A. Koestner, A.S. Leon, G.U. Liepa, K.E. Mcmartin, C.L. Mendenhall, I.C. Munro, E.J. Novotny, A.G. Renwick, S.S. Schiffman, D.L. Schomer, B.A. Shaywitz, P.A. Spiers, T.R. Tephly, J.A. Thomas and F.K. Trefz. 2002. Aspartame: Review of safety. Regul. Toxicol. Pharmacol. 35: 1-93.
7. Charu, G., P. Dhan, G. Sneh, and G. Sudha. (2012). Role of Low Calorie Sweeteners in Maintaining Dental Health. Mid-Est. J. Sci. Res.11 (3): 342-346.
8. Chadwick, W., Y. Zhou, S.S. Park, L. Wang, N. Mitchell, M.D. Stone, K.G. Becker, B. Martin and S. Maudsley. 2010. Minimal peroxide exposure of neuronal

cells induces multifaceted adaptive responses. doi: 10.1371/journal.pone.0014352.9

9. European Food Safety Authority. 2007. Neotame as a sweetener and flvor enhancer: Scientific opinion of the panel on food additives, flavourings, processing aids and materials in contact with food. DOI: 10.2903/j.efsa.2007.581

10. Food and drug administration. Health claims: Dietary noncariogenic carbohydrate sweeteners and dental caries. IN U.S., C. O. F. R. C. (Ed.).

11. Gupta, S., M. Vivek, M. Shagun, T.R. Vishal. 2012. Artificial sweeteners. J. K. Sci. 14:1-4.

12. Gardner, C. 2012. Nonnutritive sweeteners: Current use and health perspectives. Diab. Care. 35: 1798–1808.

13. Hill, A.B. 1965. The environment and disease: Association or causation?. Proc. R. Soc. Med. 58(5): 295–300.

14. Jang H.J., Z. Kokrashvili, M.J. Theodorakis, O.D. Carlson, B.J. Kim, J. Zhou, H.H. Kim, X. Xu, S.L. Chan, M. Juhaszova, M. Bernier, B. Mosinger, R.F. Margolskee, J.M. Egan. 2007. Gut-expressed gustducin and taste receptors regulate secretion of glucagonlike peptide-1. Proc. Natl. Acad. Sci. USA, 104: 15069-15074.

15. Lim, U., A.F. Amy, M. Traci, H. Patricia, M. M. Lindsay, S.S. Rachael, C. David, H.R. Albert, and S. Arthur. 2006. Consumption of aspartame-containing beverages and incidence of hematopoietic and brain malignancies.Cancer Epidemiol. Biomarkers Prev. 9:1654-1659.

16. Mejia, H. 2012 Inflammasome-mediated dysbiosis regulates progression of nafld and obesity. Nature. 482: 179-185.

17. Meister, K. 2006. Sugar substitutes and your health, New York, American Council on Science and Health.

18. Mayhew, D.A., C.P. Comer and W.W. Stargel. 2003. Food consumption and body weight changes with neotame, a new sweetener with intense taste: Differentiating effects of palatability from toxicity in dietary safety studies. Reg. Toxi. Pharm. 38: 124-43.

19. Nabors, L.O.B. 2012. Alternative sweeteners, Boca Raton London New York, CRC press, Taylor & Francis Group.

20. Olivier,B., S.H.Ahmed, C.Atlan, J. Belegaud, M. Bortolotti, C.M. Canivenc-Lavier, S. Charriere, J.-P. Girardet, S. Houdart, E. Kalonji, P. Nadaud, F. Rajas, G. Slama, and I. Margaritis. 2015. Review of the nutritional benefits and risks related to intense sweeteners. Arch. Public. Health. 73: 41-49.

21. O'donnell, K. 2005. Carbohydrates and intense sweeteners.Chemistry and technology of soft drinks and fruit juices, Hereford, UK, Blackwell Publishing Ltd.

22. Panchal, I.I., S.J. Dhrubo, S.K. Samir. 2014. Novel approach in diabetes millitus:Say no to sugar and yes to artificial sweeteners. Int. J. Pharm. Res. Bio-Sci. 3: 770-784.

23. Palese, M., T.R. Tephly. 1975. Metabolism of formate in the rat. J. Toxi. Env. Health. 1: 13-24.

24. Paul, T. 1921. Physical chemistry of food stuff. V. Degree of sweetness of sugar. In alternative sweeteners, nabors, l.O. (ed.), p. 147. Ny: Marcel dekker. Pmc specialties group, inc. Cincinnati, oh. Unpublished data.Cited by r.L. Pearson.In saccharin.Alternative sweeteners, nabors, l.O. (ed.), p. 147.Ny: Marcel dekker.

25. Qurrat-Ul-Ain, S.A. Khan. 2015.Artificial sweeteners: Safe or unsafe? J. Pak. Med. Assoc. 65: 225-227.

26. Rebaudioside, A.S. 2009.The Calorie Control Council." The Calorie Control Council | Healthy Eating & Exercise for Life. Web. 20 Dec. 2009

27. Roberts, A., A.G. Renwick, J. Sims, D.J. Snodin, D. J. 2000. Sucralose metabolism and pharmacokinetics in man. Food. Chem. and Toxicol. 38: 31-41.

28. Suez, J., K. Tal, Z. David, Z.-S. Gili, T.A. Christoph, M. Ori, I. David, Z. Niv, G. Shlomit, W. Adina, K. Yael, H. Alon, K.-G. Ilana, S. Hagit, H. Zamir, S. Eran, E. Eran. 2014. **Artificial sweeteners induce glucose intolerance by altering the gut microbiota.** Nature. DOI: 10.1038/nature13793.

29. Stonehouse, W., C.A. Cathryn, P. John, H.R. Stephen, M.A. Minihane, H. Crystal, and K. David. 2013. DHA supplementation improved both memory and reaction time in healthy young adults: A randomized controlled trial. Am. J. Clin. Nutr. 97:1134–1143.

30. Schernhammer, E.S., B.A. Kimberly, B.M. Brenda, S. Laura, W.C. Walter, and F. Dian. 2012. Consumption of artificial sweetener– and sugar-containing soda and risk of lymphoma and leukemia in men and women. Am. J. Clin. Nutr. 96**:** 1419-1428.

31. Soffritti, M., F. Belpoggi, E. Tibaldi, D.D. Esposti, M. Lauriola. 2007. Life-span exposure to low doses of aspartame beginning during prenatal life increases cancer effects in rats. Environ. Health. Perspect. 115**:** 1293-1297.

32. Spillane, W. 2006. Optimizing sweet taste in foods, 1st ed. Null: CRC, Print.

33. Shallenberger, R.S., G.G. Birch. 1975. Sugar chemistry. AVI Publishing Co,Westport.

34. Salant, A. 1972. Nonnutritive sweeteners. In: handbook of food additives (Ed. T.E. Furia), 2nd edition. p. 523. CRC Press, Cleveland, Ohlo.

35. Tandel, K.R. 2011. Sugar substitutes: Health controversy over perceived benefits. J. Pharmacol. Pharmacother. 2: 236-243.

36. Tisdel, M.O., P.O. Nees, D.L. Harris, and P.H. Derse. 1974. Long term feeding of saccharin in rats. In symposium: Sweeteners (Ed. G.E. Inglett) p. 145-158. Westport, CN: Avi Publishing Co.

37. Walters, eric d. "All about sweeteners." All about sweeteners.Web. 23 sept. 2009.

38. World Health Organization. 2008. Evaluation of certain food additives and contaminants. Forty-fist report. WHO Technical Report Series. Geneva, Switzerland.

39. Whitehouse, C.R., J. Boullata, L.A. McCauley. 2008. The potential toxicity of artifiial sweeteners. AAOHN. J. 56**:** 251-259.

40. Wade, L.G. 2006. Organic chemistry. Upper Saddle River.Pearson Prentice Hall.

Printed by Books on Demand GmbH, Norderstedt / Germany